W9-BEB-737

What You Must Do to Optimize Your Professional Digital Presence

E-HABITS

BRAND YOURSELF WITH

Strategic Social Networking
Proactive E-mail Practices
An Impressive Online Profile

ELIZABETH CHARNOCK

New York Chicago San Francisco Lisbon London Madrid Mexico City
Milan New Delhi San Juan Seoul Singapore Sydney Toronto

1 2 3 4 5 6 7 8 9 10 11 12 13 14 15   WFR/WFR   1 9 8 7 6 5 4 3 2 1 0

ISBN 978-0-07-162995-9
MHID   0-07-162995-5

This publication is designed to provide accurate and authoritative information in regard to the subject matter covered. It is sold with the understanding that neither the author nor the publisher is engaged in rendering legal, accounting, securities trading, or other professional services. If legal advice or other expert assistance is required, the services of a competent professional person should be sought.

—From a Declaration of Principles Jointly Adopted by a Committee of the American Bar Association and a Committee of Publishers and Associations

To my father, Harry Charnock, and to the wildly improbable cast of characters who together have accomplished the impossible in building our company, Cataphora

Library of Congress Cataloging-in-Publication Data

Charnock Elizabeth.
    E-habits : what you must do to optimize your professional digital presence / by Elizabeth Charnock. — 1st ed.
        p.   cm.
    Includes index.
    ISBN 978-0-07-162995-9 (alk. paper)
    1. Digital communications.   2. Digital media.   3. Work environment.
    I. Title.

TK5103.7.C54   2010
302.23'4—dc22   2010001201

McGraw-Hill books are available at special quantity discounts to use as premiums and sales promotions or for use in corporate training programs. To contact a representative, please e-mail us at bulksales@mcgraw-hill.com.

The companion software to this book may be downloaded from digitalmirrorsoftware .com, where further instructions and documentation may be found. Please note that the software will naturally evolve over time and therefore may not in every instance be identical to what is pictured in this book.

This book is printed on acid-free paper.

Contents

Acknowledgments   v

1 The Digital YOU...1

2 The Digital YOU at Work....................................13

3 Actions Speak Louder than Words......................65

4 Annoying Digital YOU Character Traits.............. 97

5 The Digital YOU in a Bad Mood.........................141

6 Love, Sex, Romance, and the Digital YOU............161

7 When the Spotlight Shines on You.....................191

8 The Twists and Turns of the Digital Grapevine.....217

9 Can the Digital YOU Improve the Real You?........237

Notes   243

Index   247

Acknowledgments

I MUST START off by thanking Will Schwalbe, who first suggested that I write this book. He also introduced me to the incomparable Eleanor Jackson, my literary agent, without whom this book certainly never would have been written.

The book comes from my experience and that of the entire Cataphora team over the eight years and counting since we founded the company. Everyone who has helped the company—employees, advisers, customers, partners, and our many supporters—has thus contributed to the book. In particular, Ron Weissman's ability to capture complex reality in a simple phrase was invaluable to me at various points along the way.

A team of Cataphora volunteers performed research and made many suggestions. Rick Janowski brought his organizational skills and humor to bear on coordinating the different aspects of the project and was indispensable to the effort. Most prominent among the volunteers were Penni Sibun, Lizzie Allen, Ken Bame, Karl-Michael Schneider, and Philip Wang. Curtis Thompson, Markus Morgenroth, and David John Burrowes produced the screenshots; Curtis was also responsible for the overall coordination of graphics and the book's website. Other contributors included Keith Schon, Raj Premkumar, Steve Roberts, Mariko Kawaguchi, Jeremy Linden, and Joshua Minor. Many people in the Cataphora engineering organization contributed to the companion software and are credited there.

I would also like to express my gratitude to Matt Welsh (and Matt Welsh), Chris Lunt, and Curtis Jackson for agreeing to let us share their stories.

I'd like to thank my editors at McGraw-Hill, Emily Carleton and Tania Loghmani, for their efforts. Under the leadership of Caroline Kawashima, our extended team at Cataphora's PR firm, Racepoint Group, was a sound and much appreciated source of advice. Thanks also to the Monaco Media Forum and all the staff at the Bagni di Pisa in Italy, where portions of this book were written.

Lastly, while it may be unoriginal, I'd like to acknowledge my friends and family for supporting me in a variety of different ways during this effort, including not seeing me for even longer stretches of time than usual.

To all of these, and others too numerous to mention, I offer my heartfelt thanks.

1

The Digital YOU

TODAY, MANY MILLIONS of people like you and me generate at least 10 different types of electronic data before our second cup of coffee. It starts from the moment we use a card key to gain access to our office and continues as we listen to our voice mail, respond to that first e-mail, and so on. Most of these actions are just minutiae, about as memorable as where we ate lunch two weeks ago. Nor are these actions widely visible the way they once were, back when office workers spent most of their days interacting directly rather than hunched over their keyboards. Yet, paradoxically, the "digital breadcrumbs" these actions leave behind can accurately capture individual behavior in more detail than ever before—not to mention permanently.

The trail doesn't stop when the workday is over. In fact, for some, that's really when the action starts. Once we're out of the office, we can tweet away to our heart's content or spend hours commenting on all of our friends' Facebook updates without fear of getting busted by the boss. Nevertheless, the second paradox of the digital breadcrumb trail is the silent, often terrifying way it demolishes the traditional boundaries between our personal and professional lives—while at the same time erecting even more impenetrable barriers.

Let me explain. Perhaps the best example of this second paradox in action is the job seeker. In many corporations, résumés land in a hiring manager's inbox only after an underling has found and attached the relevant Google, LinkedIn, and Facebook results for the applicant. All of this represents an amalgam of what the job seeker wishes potential employers to know about her, what the Internet "knows" about her, and what she allows her friends and family to know about her. The final effect is—well, unknowable. Imagine, for example, being filtered out of the set of top applicants for your dream job because you really like bowling or have a passion for collecting antique dolls. Or because a friend just couldn't resist taking a *really* unflattering picture of you and posting it on the photo-sharing site Flickr. (If this sounds preposterous, remember that when there are way too many generally qualified applicants for a given position, employers have to filter the list somehow.)

It is often said that the Internet is the most interactive medium ever. As such, if you ignore it, it will largely—if not totally—ignore you. But conversely, the more time you actively spend interacting with it—depositing different kinds of breadcrumbs as you post content, respond to blog postings, join social networking sites, and so on—the larger your profile becomes over time. The result is that some people acquire pages and pages of Google results that are really about them (as opposed to someone else who happens to have the same name), while others seem to maintain total anonymity. Thus, many of today's job seekers stumble on a new and insidious type of stature gap, one that is purely digitally driven and seems a lot like a bottomless pit if you are on the wrong side of it.

In the old days, you might buy a snazzy new suit and borrow your friend's much nicer car to go to an important interview—and, of course, spend lots of time trying to inject as much stature as possible into your résumé. But what is a coherent response to Google determining that you're obscure—or that for all intents and purposes, you don't exist at all? Is that a better fate than being deemed uncool or unenlightened, a bit too much fun, or too much of a know-it-all?

If you spend lots of time sprucing up your LinkedIn or Facebook page so you'll have a larger digital footprint with more (hopefully good and consistent) personality, but Google *still* thinks you don't really exist, do you end up looking even more pathetic and insignificant?

This dynamic gets scary pretty quickly.

The single most important thing to understand about the digital world is this: it is a place that is both enticing and dangerous in much the same way as is a foreign country in the physical world. Its seductions, such as its immense convenience, are much clearer than most of its dangers and petty mischiefs, even to the frequent traveler. Further, as we'll see, it is much safer to be a tourist than a business traveler.

My perspective on all of this is not just that of an Internet user or even a Silicon Valley technologist. I run a company that is probably unlike any other—an evidence analytics firm. When particularly large scandals, investigations, or lawsuits hit large companies, they hire us—not just for our patented software that analyzes the patterns in many millions of electronic data records, but often for our staff of mathematicians, computational linguists, fraud analysts, and other specialists who use this software to determine who's been naughty or nice.

As a result, since 2002 we've looked at hundreds of thousands of interesting e-mails and other types of electronic data and at analyses of millions more. Possibly even some of yours, dear reader! Because what few people realize is that for every glamorous lawsuit or investigation that makes it above the fold in the *Wall Street Journal*, there are at least 10 times as many that involve "normal" people—often without their knowledge.

For example, if you—or your company—are trying to sue one of our clients, and we have a lot of your e-mails, instant messages (IMs), documents, phone calls, and other data to analyze, you have reason to worry. We speak more than 20 different languages. Not only that, but our software is smart enough to know that while, for example, you may always write in Russian to Grandma, you generally don't use that language in your professional life. So if you suddenly begin, perhaps you are trying to subvert a compliance monitoring program or prying human eyes. The software will automatically detect everything from canceled meetings to missing reports to deviations from standard workflows. It will likewise reconstruct party invitation lists and determine which employees socialize in their off-hours—as well as which ones *used* to socialize but no longer do. We'll bust you for contradicting yourself, even in trivial ways.

And that's just in the first couple of days.

We also work with corporations to find any rotten apples that may be in their barrel. We work with investigators and some types of plaintiffs' firms. Sometimes we work with white-collar criminal defendants, ranging from technology executives to Deborah Jeane Palfrey (a.k.a. the D.C. Madam), to try to help establish their innocence. Or at least to help them paint a more nuanced picture of reality.

At Cataphora, we are paid to do what we like to call "corporate archaeology." This means we reconstruct often-complex real-world events on the basis of surviving data records—often millions of them. These comprise e-mail messages, IMs, documents, phone logs, and many other kinds of data records, as well as whatever relevant traces might still be had on the Internet (see Figure 1.1). But in a deeper sense, what we really do is figure out what makes the target set of people tick. Does it seem probable that Margaret could have stolen a long look at the patent draft of a competitor? *Would* she have? Did Terry lie to his investors, or was he more likely merely being careless or inept? Did Robert know that the discounted merchandise was defective, or was he probably far too drunk or hung over to have noticed? Often we examine people's actions during times of great difficulty, when everything seems to be unraveling around them. This tends

Figure 1.1 Keith Schon and Eli Amesefe, two Cataphora employees, comparing analytical results

to magnify personality traits, both good and bad, making them a bit easier to assess.

The work can be likened to reading different pieces of the cross-referenced diaries of insanely prolific diarists. However, the overall effect is vastly more potent, in part because so many real-life actions seem to be taken almost on autopilot. As such, a portrait constructed from these actions captures the essence of a person's nature with far greater clarity than anything he's consciously composed ever could. The fact that this portrait is painted with sophisticated computer programs adds the imprimatur of objectivity.

The attorneys and investigators with whom we work use our various computer-generated archaeological exercises and personal portraits in their witness interviews, depositions, and even in trials. Key witnesses and actors in a big case expect to be asked certain obvious questions relevant to that case. Their attorneys will coach them accordingly. What they don't expect is some off-kilter question about why they changed some subtle habit at a particular point in time. In other words, "When did you first start to suspect that there might be a defect in the product?" is obviously going to be a key question in a product liability case. But a question such as "Why did you stop having lunch with James on Fridays?" or "Why did you start deleting e-mails at the end of every month?" often flummoxes opposing witnesses into a perplexed stupor that benefits our clients as they cross-examine them.

After we had been doing this work for a while, we began to understand that we knew more—*really* knew more— about our targets than their spouses or closest friends did. Perhaps more than they knew about themselves. We not only knew whether they were happy or unhappy, but what made them that way and how they behaved in either state. We could observe the consistency of their viewpoints over time and whether they expressed the same opinions regardless of who was listening. We saw who was overgenerous about sharing credit and who endeavored to grab credit for anything short of inventing the Internet. We saw who was quick to lash out in anger and who was ready to forgive and forget, who was sincere and who was spiteful. We saw who did work and who mostly just complained about the work done by others. Most of all though, we saw who commanded respect, influence, and loyalty—as well as who was in no real danger of ever commanding any of these. And who had the self-awareness to know where he or she truly stood in the esteem of others.

Eventually, we came to realize something more: we were clinically capturing the character of both individuals and organizations. Really understanding the character of the key players is essential in many types of litigation, specifically those in which the facts themselves are unknown or in dispute. Of course there are myriad other reasons why such information is interesting to employers, banks and other institutions that might extend you credit, insurance companies, marketers, people you are dating, those with whom you might do business, and so on. Our job is made easier in the litigation context, because the corporations involved provide us with all of the electronic data for the relevant people, but much can be learned just on the basis of what's publicly available. The more ways that people feel almost compelled to participate on the Internet, the greater the prospects for analysis.

To go back to our résumé example, a website called Emurse urges you to use its site to ensure that only one version of your résumé is accessible on the Internet; this is designed to prevent various types of possible problems and embarrassments. For example, if different versions of your résumé are floating around out there in cyberspace, the fact that you took the liberty of slightly embellishing your title in a prior job—perhaps promoting yourself to project lead—is likelier to be noticed. Even by a person. A computer program can do this easily; it recognizes you by a combination of your name and other key information, such as details of your college degrees. Identifying a Web page or document as a résumé is also an easy trick for software, since résumés almost always contain certain common types of phrases and content, such as "education" or "work history." Detecting inconsistencies in the dates of employment or the phrasing of a title associated with a particular job in different versions of the résumé is only slightly more difficult; the same is true for catching verb "upgrades," such as "participated in a team of 20" versus "led a team of 20."

Extending this example a bit further, imagine a résumé that has been upgraded several times over a period of months. The perpetrator now looks like a desperate liar and will likely be left with both the time and the need to upgrade quite a few more times as a result. Likewise, several versions of a résumé that have each been tailored to appeal to a specific employer or type of employer can easily seem obsequious and insincere in this context, even though such tailoring is often encouraged by career coaches.

This is just one narrow example, however. Many others spring to mind—evolving profiles on dating sites, finance-related newsgroups, and other forums that over time can expose you as having been cosmically wrong, such as when you confidently declared that it was obvious that a particular presidential candidate would win the general election in a landslide only to have the candidate wipe out early in the primaries.

Tack all of the personal digital breadcrumb–generating activity on to the literally billions of e-mails, IMs, and other forms of electronic communication that occur every day during the workweek, and you have millions of digital portraits capturing every aspect of people's lives. Any current employer who wishes to do so can easily amass much or all of this information; in the United States, any data you create on your employer's dime and time belongs to them. In other words, privacy really doesn't apply when you are using your employer's computers and network.[1] As the percentage of the workday spent online continues to grow, more and more employers will find themselves in the position of needing to take further steps to monitor their employees. The motivations are myriad: preventing fraud, reducing some kinds of insurance premiums, determining who is productive, or simply better understanding the actual functioning of their business.

What can be known about you from your digital breadcrumb trail, and how? Let's start off with a simple example. The data visualization in Figure 1.2, which is from the free Digital Mirror software available at digitalmirrorsoftware.com, illustrates several possible ways of assessing the importance that different individuals assign to one another based not on any company organizational chart or social network links proclaiming eternal friendship, but solely on empirical assessment of behavior. The purpose is to identify relative differences between how one person treats the individuals he interacts with the most and how these same people treat him.

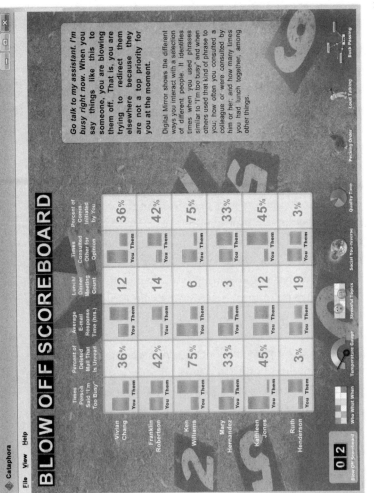

Figure 1.2 Blow Off Scoreboard report (*Background photo courtesy of Ed Sanders—http://*
bit.ly/abRmCl)

Some of these categories require merely tallying data records (like counting up the number of IMs or phone calls exchanged), while others require the application of computational linguistics techniques. For example, accurately detecting an "I'm too busy" response requires automatically determining the difference between someone saying, "Things are going great, but I'm really busy," and "I don't have time to go out to lunch with you because I'm too busy." Determining when someone is soliciting an opinion is also a bit involved; we'll come back to this in a later chapter. For now though, the main point to note is that we try never to draw a conclusion based solely on any one kind of measurement. Reality is complicated; multiple probes—and multiple types of probes—of the same question are necessary to get a reasonably correct answer for most things.

This type of coldhearted, objective analysis underscores the fact that none of us treats all of the actors in our professional or personal lives equally. Not even close, in fact. We eagerly grab the chance to interact with some, while striving to duck out of others' sight lines—just as other people may do to us. Still, when people see this kind of analysis done on themselves, a common reaction is that they can't believe they were so consistently rude to this or that colleague. "Surely there's some mistake," they usually say.

No, there isn't.

It is simply a matter of not having previously assessed reality in these terms. Most people tend to think in the old-fashioned way, at the unit level of the individual phone call or e-mail. They may know they did not return a particular phone call as quickly as they might have. But they don't think of it in quantitative terms; for example, how much faster they typically respond to Suzy than to Joan. Likewise, while you may realize that you tend to brush off an overly talkative coworker when you are especially pressed, you may not realize that you nearly *always* do it. Or that you sometimes leave behind a mile-wide digital breadcrumb trail on a corporate wiki—or anywhere else the talker may be able to see it—just after you've told him or her you are headed into a meeting for the rest of the day.

So what is the Digital YOU then? It is much more than the sum of the breadcrumbs you leave behind in your travels through cyberspace. The Digital YOU is a complex mosaic of habit, subconscious acts of both omission and commission, and premeditated presentations. It is how your peers and friends, bosses and family actually experience you, as more and more of your life in the real world takes place in the digital one. The purpose of this book and the associated software that you can download from our website is to help you begin to experience your Digital YOU in the same way everyone around you does.

————

A final note: We work under the umbrella of attorney-client privilege, just like the lawyers on TV. So we can only name names when our clients give us permission to do so, and most cannot, because the matters in question are still pending as of this writing. However, all of the various anecdotes included in this book come from either a case in which we were involved or a case that one of the law firms with which we regularly work contributed. In a few cases, we used something we read about in the newspaper that was too good not to include.

The software industry stories all come from the direct experience of someone at Cataphora or one of our friends.

2

The Digital YOU
at Work

IT IS PROBABLY best to begin by explaining a bit more about what we do at Cataphora and how we have come to have this unusual view into the strange and complex ecosystems of Digital YOUs that are modern corporations. Historically, most corporations have cared much more about measuring the results of professional workers than analyzing the behaviors that led to those results. However, as both individuals and the content they create become increasingly intertwined, it becomes nearly impossible to separate the understanding of results from the understanding of behavior. In fact, understanding *why* things happen is becoming more and more important in society in general. In our increasingly litigious culture, hundreds of millions of dollars can hang on the "why"—on whether something appears to be a tragic mistake or an isolated exception as opposed to a cynical, routine circumvention of decency or accepted practice. Also, on the most basic level, the causes of things that happened used to be fairly obvious and therefore less interesting. This is no longer the case.

Consider that 30 years or so ago, most office workers had typewriters rather than computers. The idea of working your

way up from the mail room existed because large corporations all had to have a significant number of junior employees to sift through and physically distribute interoffice and
other mail. Working with colleagues usually meant being in
the same room. Certainly you could call people, but in the
days before cell phones, your chances of reaching someone
directly on the first or even second try were nowhere close to
what they are today.

If you knew that one of your coworkers had created a
really great presentation or report on a particular topic, you
could borrow relevant pieces for whatever you were working
on, but that would usually require rekeying, word by word,
the content in question. Select-Copy-Paste was still the stuff
of science fiction. And if you wanted to find some great presentation that you hoped was merely hiding in some corporate nook or cranny, you would have to rely on either the
assistance of an old-fashioned corporate librarian (think
Katharine Hepburn in *Desk Set*) or the collective memory of
colleagues. This would probably take quite a while, because
you'd have to go from person to person until you hit pay
dirt. By contrast, people nowadays complain about having
to visit multiple portal sites to find exactly what they are
looking for.

All things considered, you were largely on your own back
then. Both helpful human interaction and reusable content
were much harder to come by and hence much more limited
in scope than they are now. As a result, the world was a relatively simple place.

Now fast-forward to the present day, where complex multimedia presentations and documents assembled largely
from existing building blocks are the norm. Many more
people and much more information can now easily contrib-

ute to—and exert influence over—any decision-making process, creative endeavor, or other activity in the workplace. But by the same token, a great deal can just as easily disappear—or be made to disappear—unnoticed amid the chaos. Let's say you found—in your e-mail or on the corporate wiki—the presentation given in the big pitch meeting to an important new client. It includes most of the relevant points but seems to omit some. You wonder why. Well, there could be many reasons for it. Let's look at the most likely:

- ◆ Laziness or unawareness on the part of the author
- ◆ A desire to keep the presentation short and pithy
- ◆ An honest belief on the part of the author that the "missing" content was incorrect in some way
- ◆ Political motivation to gloss over certain issues
- ◆ An active desire to conceal the truth or to commit some kind of fraud

And that's where we come in, with our corporate archaeology tools, to determine—based on the remaining electronic data artifacts—what probably transpired: the why.

I say "probably" because what we do is play the probabilities, just as good lawyers or detectives do. To do this, we use technology and our experience in applying that technology to real-world situations, but we also rely on a very fundamental truth: character is destiny.

When I was a freshman at the University of Michigan, one of my professors told us that that phrase, attributed to the ancient Greek philosopher Heracleitus around 500 B.C., was the most important thing to take away from our whole college education, because so much else followed directly from it. He was right.

If there is one thing we've learned since we started Cataphora, it's that if you can truly understand someone's character, it's usually not very difficult to determine what they did or did not do, or to dismiss a number of theoretically possible scenarios as unlikely. The hard part is really understanding character. After all, most people want to be seen as nice and reasonable—even those who, in reality, are mean-spirited and short-tempered. A certain amount of diplomacy and even subterfuge is a necessary survival skill, particularly in a corporate environment. Showing your true nature can be dangerous. This makes detecting the truth a significant challenge.

But we have a unique advantage. Rarely do we actually meet our "targets." Instead, we study a comprehensive set of their electronic artifacts, sometimes dating back years. This means e-mails and instant messages, documents and Web pages, calendar appointments and phone records—indeed, anything that still survives. For most people, this amounts to hundreds of thousands of data items, mostly short messages that collectively document the day-to-day life of the individual in question.

Most people assume that our job is to find the smoking gun, as forensics experts do on TV. However, just the reverse is true, and this is what makes us necessary. In fact, the way I often explain it is that in the unlikely event that the following e-mail exists in a fraud investigation, our assistance would not be needed:

> I know I shouldn't have done it, but when the CFO discovered that I had embezzled the $300,000, I had no choice but to kill him and hide the body in one of those large canisters that were being hauled away to

the city dump. He was a real bastard anyway, and no one—not even his wife or dog—will miss him.

Such true smoking guns are exceedingly rare in real life for all kinds of reasons, chief among them that most executives are given pretty clear instructions about what types of things should never, ever be put in writing. However, virtually everyone has put *something* in writing at one time or another that, if widely known, would be embarrassing at best and could cause life-altering problems at worst. As we'll see in more detail later, common examples include admitting to different types of cheating on (or off) the job, poking fun at coworkers or superiors in ways that might look particularly insensitive or unenlightened, and the expression of a wide range of ill-advised fantasies.

If we look hard enough, pretty much everyone has said something that would have been best left unsaid—and certainly best not committed to permanent digital form. Because such (hopefully) brief lapses in judgment are so common, they may be interesting from our perspective, but they generally don't tell us anything except that the people we are studying are human. This means that on some days they will (1) wish aloud that they had married someone who was 20 pounds thinner (like maybe that hot new secretary); (2) not resist the temptation to insert e-graffiti on a photo of their boss; or (3) write an e-mail they never intended to actually send but sure let off some steam.

Important note: Employers don't necessarily share this view, and many people are fired each year for precisely these types of indiscretions. However, in our work, we are generally focused on reconstructing the relevant series of events in an investigation; other unrelated, injudicious acts someone may have committed are only of interest insofar as they help us understand that person's probable role in the events we do care about. Our liberal view is also a necessary by-product of what we do for a living. New hires at Cataphora will often hear things like "Oh, you think *that's* bad? Well, here are a dozen recent examples that are much worse."

By contrast, human resources and legal departments tend to see such bad documents only when complaints have been made and therefore, not irrationally, tend to regard the authors of said documents as "perps" who need to be shown the door as quickly as possible. Because we are looking at broad cross sections of people against whom no complaint has been filed (or ever will be), we have a much more nuanced and empirical view of normality.

So if we do not rely on smoking guns to build our profiles of Digital YOUs, what do we use? The answer is a comprehensive model of all the small, routine tasks that people do day in and day out. This is ultimately the only deep truth. Just as the list you penned of "10 fantasy ways to get fired" probably wasn't your finest moment, the important presentation you polished to perfection for a month is also not an accurate reflection of you. Both may reflect your creative prowess, but they also represent the extreme opposites of your professional spectrum. Most of what happens in the workplace falls somewhere in the middle.

As such, we care about things that may seem abstract on the surface:

- ◆ How consistently do people do the things they do? How flexible are they in their working habits?
- ◆ What kinds of things seem to upset them? What do they usually do when they're upset or angry?

- Conversely, what things seem to really engage them?
- To what extent do they maintain their opinions? Do they frequently contradict themselves if it seems expedient to do so or refuse to admit the possibility of being wrong under any circumstances?
- What kinds of communication styles do people have—formal, didactic, chummy, peppy? Is it the same for everyone with whom they interact?
- How carefully do they manage their personal data?
- Are they typically cheerful or grumpy? Bitingly sarcastic or naively optimistic?
- What kinds of relationships do they tend to have with their coworkers? Are they admired, merely tolerated, or despised? Whom do they admire?
- How stable are their workplace relationships? What characteristics do people close to them seem to share?

We are fond of saying that context is everything. For our purposes, behavior is a complex set of actions and reactions. For example, if we happen upon a "10 fantasy ways to get fired" list at the beginning of an investigation, we know there are two possibilities:

- The author is a jokester looking for attention in a passive-aggressive way and has a well-established pattern of trying to get it by breaking people out of the daily doldrums with over-the-top, don't-let-HR-see-it jokes targeted at some part of the work environment.
- Something has happened to really piss off the author, because the behavior is completely inconsistent with his normal actions.

From our point of view, these two things couldn't be more different. Our interest in the first person is close to nil, but we would immediately focus on the second. What happened to get that person so angry? Is it related to something we care about? Is he a possible whistle-blower? Is he a reliable witness, or does he have an axe to grind against someone—and if so, who? Yet it is the very same list in both cases. What makes it different is the context. The guy who is insecure and seeking attention has probably been this way for much of his life. So if he hasn't done anything really bad before, other than sending out tasteless or tired jokes, he is unlikely to do so now. In short, he may be unattractive, but he is unlikely to be anything worse. However, the guy who may seem to be a straight arrow but believes he has been wronged in some major way can become a significant wild card. In our world, this can mean anything from absconding with sensitive and highly valuable company information, to testifying against our client, to committing various types of fraud (or any number of other bad things). Alternately, he may be angry because he has seen something that horrifies him, and he knows something that our client's lawyers need to know before the other side does.

This example is a simple one; as you'll see, most are more complicated or subtle. Nevertheless, the fundamental principle remains the same: people are creatures of habit, especially in an organizational context. This means that if you can capture what "normal" is, you have largely won the battle, because you can identify the abnormal. And while not every abnormal thing is bad, the vast majority of really bad things are almost always, by definition, abnormal. (Contrary to common belief, if everyone robbed and pillaged on the job, few corporations would still be standing. The reality is that the vast majority of people will do little more than cheat slightly on a travel expense report or pocket a nice pen, if that. But even a small number of well-positioned bad actors can wreak havoc on the biggest corporation.)

That said, the Digital YOU is much more likely to be caught lying, cheating, and stealing in the workplace than the *real* you.

For most of us, the lies are small and white, aimed at avoiding something we deem both unnecessary and unpleasant, such as lunch with an especially tedious colleague. The cheating may involve writing a report that suggests you performed more research than you actually did in order to get to the obvious and predestined conclusion. The theft may be as simple as silently "borrowing" some content that perfectly captures the point you were trying to make.

While these examples may all be small peccadilloes in the broader scheme of things, the simple fact is that many sins, great and small, are much easier to commit while hiding behind a computer screen. For example, you don't even have to tell the tedious colleague you are not available for lunch; you can just change your Skype status to "Away." And there's no need to look down at your shoes as you copy and paste a list of links to websites you used to perform the extensive research for your report, even if the list comes from the one and only site you actually visited. (Be aware that we can bust you on this kind of thing easily enough, as you'll see when we get to the section on automated employee performance analysis.) The scale of possible sins may differ widely, but the psychology and motivations remain the same: to gain stature, popularity, or respect; to more easily achieve a desired goal; or to avoid some type of unpleasantness.

Evasion in particular is vastly easier in the digital world, where the closest thing to seeing someone in the hallway or in front of the coffee machine is seeing that your status is "logged in" on Skype, Yahoo!, or the equivalent. Even then, there are always ready excuses for why you ignored someone's attempts to contact you, many of which are impossible to disprove even if they are unlikely. You were concentrating hard on finishing something important. Someone else was using your computer. Your kitten was rolling around on your keyboard. You accidentally left the stereo on loud before going out for a run, and the vibrations caused your mouse to jump, so you appeared to be available when you actually weren't. And so on. By contrast, if a coworker—or, worse still, your boss—sees you in the physical world, wandering around and casually chatting to people as a major deadline looms, it can backfire even if you really were just taking a legitimate break after too many hours glued to your monitor.

In fact, the Digital YOU can be much more slippery than the real you can. One particularly elegant example of this comes from a study performed at DePaul, Rutgers, and Lehigh universities.[1] Forty-eight M.B.A. students were given $89 to split with an unknown person they were to contact in writing. Students sending a handwritten note lied about the total sum of money 64 percent of the time. But students sending e-mail to their partners lied about the amount more than 92 percent of the time. A second test found that the rate of lying remained the same even when subjects knew their partners.

The Digital YOU is therefore harder to manage in the workplace. If you work in an office environment, you probably spend a good chunk of your day e-mailing, sending

IMs, and surfing the Web. Most, though certainly not all, of it is work-related. But the amount of personal business you do is now much easier to conceal than it was when the only means of immediate communication you had at your desk was the phone. Back then, if you were spending lots of time during work hours in detailed consultations with your mechanic or your personal trainer, everyone within earshot knew it. Not so with e-mail and instant messaging. Since so much content is readily available from so many different sources, it can now be much more difficult to determine who is actually spending lots of time creating content—or at least adding true value by carefully selecting existing content of the highest quality and relevance—and who is indiscriminately swiping it from anywhere they can so as to have more time to spend IMing the hottie they met at the bar the other night.

There's also the somewhat freakish dynamic that I like to call the Dorian Gray effect. In the novel *The Picture of Dorian Gray* by Oscar Wilde, the protagonist sells his soul for eternal youth; his portrait reflects his true age while he remains physically unchanged. Over time, the portrait grows more and more hideous, reflecting his cruelty and debauchery, while the real Dorian Gray retains a young and innocent appearance. He gazes happily at his handsome, unchanging reflection in the mirror while hiding the increasingly unappetizing portrait.

A similar divergence often happens between the meticulously groomed digital personas people construct for themselves on social networking sites and the portrait of their Digital YOU constructed from pulling together all available bits of data from all available sources that relate to them. This is because often the more time and effort people spend obsessively maintaining and polishing their digital image, the less time they have to spend doing actual work. It is only a matter of time before this fact is widely noted and resented by others, causing related artifacts to visibly litter the digital landscape. As we'll see in this chapter, in this situation the full Digital YOU portrait becomes gradually less attractive over time—it can easily hit the hideous mark and keep right on going.

We often visualize this phenomenon by means of an old-fashioned meter representation, such as the one in Figure 2.1, but in this case the meter indicates the observable opinion that individuals or groups have of a person with whom they regularly interact. As this visualization is "played" over time, the needle will drift increasingly to the right if the topic is a colleague who spends far more time self-promoting than doing. (Figure 2.1 is a visualization from the Digital Mirror software that contrasts your opinion on different topics with the opinions of those around you.)

In situations where separating the different sources of opinion is unnecessary, a fun way to visualize the collective opinion the group holds of a particular individual is to start off with a line drawing of a face that is expressionless. As data about the person is collected, the expression will begin to reflect the nature of the data content. For example, the detection of enough slightly negative content will cause the corners of the mouth to turn down slightly. Lots of decidedly negative content emanating from different sources (remember, no one gets along perfectly with everyone) will turn the corners of the mouth down in a frown, until at some point, it becomes a grotesque-looking grimace. Conversely, if there is significant praise for the person, the mouth will form a smile. This type of visualization approach works well when

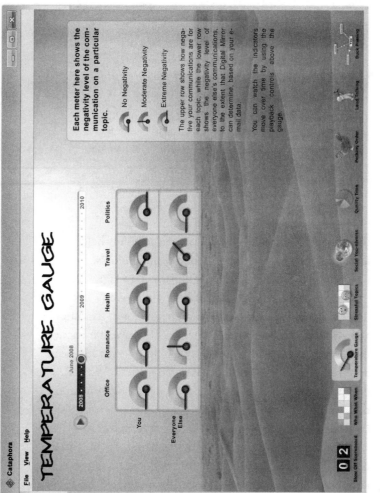

Figure 2.1 Temperature Gauge view (Background photo courtesy of J. Gremillot—http://bit.ly/94HZEB)

| David | Dorothy | Tim | Christina | Gabe |
| Mooney | Davis | Wirtz | Polanco | Franklin |

| Katie | Lee | Heather | Ron |
| Nolan | Liao | Evans | Wright |

Figure 2.2 "Wisdom of the collective" management team portrait

the goal is to compare and contrast the generally held opinion of a similarly positioned group of individuals, such as the executive management team illustrated in Figure 2.2.

This notion of the wisdom of the collective is nothing new. It has always existed in large companies. Traditionally, some executives have merited hushed awe at a chance meeting in the cafeteria, while others have generally been despised as spineless or otherwise incompetent. Why? People "just knew." The key difference now is that commentary that was once reserved for whispers around the water cooler is often memorialized in some electronic form.

One common example of this dynamic is an e-mail originally sent by an executive who is not seen as particularly credible. The e-mail is helpfully annotated by some of its recipients to indicate which parts are true, which are not-so-true, and which are belly-laugh-funny-untrue. Or perhaps sections of a blog posting by a high-profile executive are copied and pasted into a new post with similarly helpful annotations. If you get people angry enough, they may even spoof your blog, if it is a highly visible one. This happened to Jonathan Schwartz, the former CEO of Sun Microsystems.[2] The following excerpts are from an actual post on the Yahoo! Finance message board for Sun at about the time he had the dubious honor of receiving a 24 percent approval rating on the website glassdoor.com, a site offering employees the opportunity to post anonymous reviews of their employment conditions. The post was titled "Jonathan Blog—update page 5":

[Entry 2009, 08 March]
Oh boy, I suckered those idiots over at Wikipedia to put all that nonsense on MySQL for free. What a bunch of idiots. Then I played hard to get when it came to support, so they tossed out all the Sun hardware one afternoon about a week ago. Turned around and put in that HP stuff with Linux. But that's OK, I'll make a press announcement about it and make it look good. I'm going to have all those Wikipedia Republicans fired when I get back to the office. One thing I do well is fire people.

You little investors started all of this. Now they won't talk to me over at Southeastern about my golden parachute thingies. You wait. We'll find out who you are.

[Entry 2009, 09 March]
And speaking of SouthEastern, who ARE these guys? I mean, Sun is MY company right? I own the entire board, the board are my little bitch slaves. **I** am the one that came up with the free software idea. **I** am

the one that came up with the fire-them-all idea. And
I am the one that chitcanned all the hardware so
we could give away more free software now wasn't I?
I'm the one that hired 10k people in the last 2 years,
and I'm the one that fired them all this month too. So
WHO are these guys to tell me I'm not going to get my
$348 mil severance bonus? Who are these guys?

That big ugly idiot over there at SE insulted me
when he told me if I left the company right now, he
would give me $1 mil in cash and 3 rides on his jet.
But—I had to leave before the end of March. I pleaded
with him, in that my work at Sun is not done—I still
have more than 6,500 people to fire, and my hair will
need to be washed and rebraided sometime this
month also. This can take days.

When such content is on the Internet, commentators can
annotate freely, generally without fear of detection, so long
as they don't do the dirty deed from within the corporate
network and keep their commentary to interpreting widely
available information.

While much of this morphing of the beautiful portrait
into something less attractive happens behind the scenes in
various private communications, the truth is you don't need
to see everyone else's private IMs and e-mails to observe the
phenomenon. Yet few people seem to see anything that con-
flicts with their idealized view of their own Digital YOU. To
a large extent, this is simply an aspect of human nature that
remains unchanged from the physical to the digital world.
People often become more attached to a mirage than to reality; after a while, they tend to forget that the oversized fish
they're holding up proudly in a picture was actually caught

by someone else. That is why most people, when they gaze into the digital mirror, see only what they put out.

A favorite and easy-to-understand example of this is the person who prominently boasts certain specific types of expertise on every personal page but is demonstrably the least likely person in the organization to ever actually be consulted in any of these areas.

All conversation on the particular topics of interest can be represented as lines connecting the individuals engaged in the discussion. To the extent that some people converse much more frequently than others on the topics in question, the software will automatically draw such people closer together, with occasional participants being drawn out toward the edges.

Computer scientists call this type of figure a graph. A very simple example of a graph may be seen in Figure 2.3. In each graph, rectangles represent individual people and are connected by lines to indicate some kind of real-world connections between those people. The specific type of connection varies based on the purpose of the particular graph. In investigative usage, in which data from hundreds or even thousands of interrelated people may be analyzed, graphs containing hundreds of thousands or even millions of people are constructed. The explosion in numbers happens because each person for whom data is collected has hundreds or even thousands of people with whom he or she interacts. (If you wish to see examples of real-world graphs, many are to be found at digitalmirrorsoftware.com.) Many of the types of graphs that we use in our work are calculated to reflect the important social network concept of *centrality*. That is, as noted above, it makes a difference whether a particular rectangle is placed near the center of the graph or out

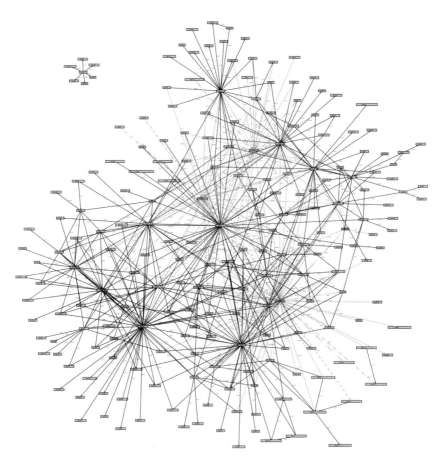

Figure 2.3 Example, small graph

on the periphery. Coloring may also be used to help identify the most central people in a densely interconnected graph. Graphs such as these illustrate that some people are always at the center of any social activity within a particular group, while others are bound to be at the center of any discussion about a particular topic. Conversely, other people may be out in the boondocks or barely participate at all. If someone who is well out of the visual center claims to be the in-house expert on the topic in question, he is either badly delusional or trying to con you. Either way, it is someone you want to avoid. When there actually is a single universally recognized expert on an important topic, the resulting graph will somewhat resemble a bicycle wheel, with the expert at the center. It is difficult to game any reasonable measure of centrality by simply sending lots of e-mails about a given topic or similar methods, because interaction is a key part of the calculation and sending out lots of messages that are largely ignored and never acted on will not help your score. For example, if I send everyone in my work group lots of invitations to parties, lots of instant messages, and so on, but no one replies, that generally puts me even farther from the center than if I did nothing at all. Simply put, introverted is one thing; pathetic is something else.

If a person really is regarded as the expert she proclaims herself to be, she should be noticeably more central in graphs showing topics related to her expertise than in those related to other types of interactions. This means different people frequently and repeatedly ask for her opinion; she is invited to meetings where these topics are a major part of the agenda; and she generates content that is commonly downloaded, copied, and referenced. Given that computer programs have long been doing a reasonable job of automatically detecting the presence of specific topical content, it is not difficult to amass such data. (Such programs are far from perfect, but perfection is not necessary to generate a ballpark understanding of a situation. Further, when performed by people, topical categorization is often highly subjective, so it is generally not clear whether machines outperform people or vice versa.)

Professional puffery is a common sin that is seductively easy for the Digital YOU to commit. Many community sites are even helpful enough to provide a list of areas in which you could be an expert; all you need to do is check the ones that apply. Unlike the painstaking care you would normally take to craft a résumé, this is so casual and quick that it is easy to forget you've done it. But once you've asserted that you are an expert in something that others notice and respond to, it becomes a bit like that oversized fish in the photograph that really was caught by someone else. It makes you more attractive and thus becomes hard to part with. Perhaps some people who start off this way actually do become experts in whatever area it is. But most sink quietly into a pleasant cloud of delusion, at least until a large round of layoffs hits. At any rate, if your employer happens to be a customer of ours, your Digital YOU's claims may very well be checked against reality at some point.

This self-aggrandizing view is not merely a question of human nature. It is also a side effect of how Google and other major portals on the Internet work. There is also the fact that most corporate data (like an HR database in which you more carefully considered which primary areas of expertise to note for yourself) isn't available to the average employee for a wide variety of reasons, ranging from privacy and confidentiality issues to potential liability.

Let's take a concrete example. If you Google my name, Elizabeth Charnock, as of this writing, the first page of results is all about me rather than other people who might have the same name. Many of the sites at the top of the list are big hubs in the digital universe, because that's the essential strategy Google uses to separate the digital wheat from the chaff. Major media appearances tend to float to the top.

Since we have a good PR firm, this means you'll see lots of nice, impressive references to me. If you're me, that's a good thing. Now let's say you don't happen to like me for whatever reason. (Perhaps I rejected your résumé out of hand because I found one of your hobbies to be unattractive.) Good news for me: it is difficult for you to do much to me on Google—unless you hate me enough and have the resources to engage a PR firm that is at least as good as mine to do so. Which is not very likely.

However, if I have accrued a number of enemies over the years in the real world, marks of resentment and hatred are left in the digital world. They may be easy to dismiss, because they are several "next" clicks away. But they are there nevertheless, lurking in the digital shadows. No individual one may rise to any particular level of significance. But the question is, how many are there? Do they all seem to hate me for the same reasons or for different ones? (And if they have the exact same reason, does it seem likely that their multiple digital identities actually map to the same person?) How do I compare in this regard to other CEOs of comparably sized companies? These are exactly the sorts of questions we ask ourselves during investigations. We know there is rarely a simple truth waiting to be discovered. Somewhere out there, there's even someone who hates Mother Teresa.

This is where the Dorian Gray effect comes in. The longer and larger my digital presence is, the more pronounced the "aging effect," regardless of what I do (so long as I am perceived to have any modicum of power in the virtual world or the real one). But as with real-world aging, the extent varies dramatically. The wittiest suggestion I've heard for dealing with your digital detractors is to co-opt their identities; for example, by appropriating someone's typical user name on a social networking or other site where it's still available and then posting comments under that alias, thus making the person come off as the ranting loon he clearly is. Operating along these lines, I might post something like the following—under the appropriate pen name of course:

> It is because Elizabeth fired me with no warning that I didn't have the opportunity to grab the pictures from my laptop of the three-headed space aliens from that party in Vegas that could have set me up for good. She really wrecked my life.

All of this matters because there is no question that Internet portals like Google and LinkedIn are ubiquitous in a professional context, and they therefore contribute significantly to perceptions of the Digital YOU in the workplace. For example, years ago, if you were invited to attend a meeting that included some coworkers with whom you were unfamiliar, you'd look them up in an organizational chart. That would allow you to understand the basics, like who they reported to, what their titles were, what groups they were in, and so on. You could determine their name, rank, and serial number, but nothing more. Now you merely Google them— or perhaps go directly to LinkedIn or a similar site. Not only can you get their titles (which in many companies provide most of the "official" information you really care about), but you can get some idea of who they run around with, how they've spent their time in the digital world, what others say about them, and so forth.

This raises the twin issues of *personal digital brand management*—how you maintain and groom the Digital YOU—

and *digital identity integrity*—essentially, ambiguities or errors in associating the Digital YOU with the real one. You may think the latter isn't a real problem, because even though there are surely other people out there with the same name as you, they probably have different occupations, live in different cities, or for whatever reason are not likely to be readily confused with you.

Think again. What technical people refer to as "collisions in the namespace," meaning the same name being attached to multiple people, is happening more and more. Essentially, the more you put yourself online, the greater the probability of such collisions, especially if you don't do smart things like always using both a first and middle name or including an unusual nickname to distinguish yourself from others with the same name.

The following unbelievable—but completely true—story is an example of just such a collision and the havoc it can wreak.

The Tale of Two Matt Welshes

Flash back to 1995, when Google didn't even exist and most of the people who were truly prolific on the Internet were in the computer industry. Two of these people were named Matt Welsh.

That's not a terribly uncommon name, but that was only the beginning of the similarities. Both men were computer scientists working in related technical fields and living in or near San Francisco and Silicon Valley, the heart of the computer industry. Both had similar undergraduate degrees, both generated a fair amount of content on the Internet, and both had an established presence there. Indeed, at one time, the domain mattwelsh.com belonged to one, and matt

-welsh.com belonged to the other. Both men contributed large amounts of content to online computer science newsgroups; for example, they responded to many technical questions posed by others. Because—for obvious reasons—the two Matt Welshes were so often confused with one another, they linked their personal websites to each other to help route confused cybertravelers to the correct place.

What makes the story noteworthy is that one of the Matt Welshes was also aggressively using the Internet to advertise his successful side business: a stripping, exotic dancing, and nude modeling business that included an appearance in *Playgirl*. This content was, of course, not intermingled with his technical content, making the confusion with the other Matt Welsh that much more likely. As of this writing, Matt's modeling portfolio and related content are no longer publicly available on the Web. At some point in the late nineties, he determined that his digital footprint was simply too large; he was drowning in e-mail from fans and could no longer find the time to respond to even a fraction of them. As a result, he took his website underground. (If you try to access his old website now, you'll get a Web page that says only, "There's nothing to see here.")

The other Matt Welsh, who as of this writing is a computer science professor at Harvard and writes books on the Linux operating system, had a sense of humor about the numerous confusions. Indeed, the two real-world Matts became friendly over the years. At one point, the professor even Photoshopped pictures of himself with the head or torso of the other Matt on his website as a joke. In return, Matt the model placed the following linked text on his home page: "If you're looking for the Matt Welsh who is a Linux god, I'm not him, but you can find him here."

While this mix-up caused no more harm than some momentary embarrassment for those who may have electronically propositioned the professor, it is easy to imagine the negative repercussions it could have had for one or both of the Matts. Harvard might have thought twice about hiring a stripper as a professor, and someone considering the other Matt's services for a bachelorette party might have been dissuaded if she had made the reverse mistake.

This story is merely an extreme version of something that happens every day. What's important to understand is that it almost never happens intentionally. Matt the stripper wasn't trying to inconvenience, confuse, or embarrass anyone. He was merely trying to use the Internet to promote his business, which he did—very successfully. It is a reminder that it is a difficult transition from a world in which your name and street address are sufficient to uniquely identify you at the post office to one in which people expect to be able to find anyone they want—anywhere in the world—within minutes, based solely on a name and one or two basic pieces of data.

As the preceding story illustrates, if you didn't know what each Matt Welsh looked like in the real world and didn't have any information about either of them beyond that they were software engineers in the San Francisco Bay Area, it would be incredibly easy to confuse the two. Many people did. In the real world though, such confusion would be highly improbable. It's a safe bet you can correctly match the photographs in Figures 2.4 and 2.5 to the correct Matts.

This is a new type of confusion inherent to the digital world, since in the real one, such confusion could only occur if the two Matts had, at one point, also had the same or very

Figure 2.4 Matt Welsh, software engineer/male model

Figure 2.5 Matt Welsh, software engineer/Harvard professor

similar street addresses—the rest of the similarities simply wouldn't matter. It's worth looking into, because if you have a namesake with whom you could be confused, you have at least two possible and significant dangers to contend with: (1) he could do things that a potential boss, in-law, college admissions officer, and so on might find offensive; and (2) the confusion could somehow open the door to disappointment, such as when you walk into a room looking like the only kind of agent you've ever had was of the State Farm variety. In both cases, the existence of a namesake might greatly impact your real life without you ever being any the wiser.

Of course, the ideal situation is to have your name be the dominant one in the global namespace, as mine is. I'm a bad example though, because (1) I'm in the computer industry; (2) I'm a CEO; (3) I'm an author of a book; and (4) Elizabeth Charnock is not a common name. Failing this type of happy combination of facts that enables such domination, you can hope that your namesakes are obviously different enough from you so as to make confusion unlikely. Of course, as the title of a famous book on selling says, hope is not a strategy. It is a good idea to look at Internet hub sites like Google or LinkedIn from time to time to spot possible sources of confusion. At best, knowing that you may be confused with other people allows you to figure out how best to deflect this confusion in the situations that are most important to you. If the worst case happens, a proactive, mutual, good neighbor strategy deployed with a deftly humorous touch, such as the one employed by the two Matts, is the best approach.

Unless you get extremely unlucky, like Silicon Valley engineering manager Curtis Jackson, a once-a-month check should be all you need. Like many people in the valley, Curtis made his résumé—with his cell phone number on it— available on his home page at curtisjackson.com. However, this was before the rapper known as 50 Cent gained widespread popularity. Crazed fans trolled the Internet looking for some way to bypass all the other groupies. Since 50 Cent's real name is (you guessed it) Curtis Jackson, a good number of these fans stumbled on the engineering manager. Many were not deterred by the myriad clues that clearly suggested the proprietor of the site was unlikely to be their beloved "Fitty." So they sent letters and e-mails—and, of course, called at all hours of the day and night. Despite many countermeasures, the only thing that even put a dent in the amount of fan contact Curtis received was the eventual decline of 50 Cent's popularity.

How to Protect Yourself at Work

Highly improbable doppelgangers and detractors crouching in the barely visible tiers of Google results are far less of a hazard for most of us than are the people with whom we have to deal digitally every day on the job.

If you've worked in corporate America for any length of time, you've likely experienced at least one near-career-death incident involving digital media. Most likely, you were the victim of a wanton and perhaps even malicious bcc'er or compulsive e-mail forwarder seeking entertainment to relieve the office doldrums. Maybe you accidentally left some indiscrete speaker notes or other annotations in a document noting just how much you had to dumb down this presentation for management or a particularly dense customer. Or you didn't realize that the compliance department was closely monitoring all communications of a colleague against whom some kind of complaint had been made. At least, not until you too were sent to sensitivity training.

We all have private opinions, fears, and insecurities, regardless of the company's official position or party line. Often these sentiments are widely shared by those around us, which emboldens us to express them. And these days, most of us are never far from a device that, with a few keystrokes, allows us both to let off steam and to bond with fellow sufferers—causing potentially massive fallout in the process.

Consider this: the more steam you let off, the truer the sentiment being expressed; the sharper your wit, the more widely your message is likely to be forwarded in a chain reaction, ending up who knows where. Maybe this is how you discover that one of your colleagues really doesn't like you—she forwards your message to the HR or compliance department. Or you may find out just how effective and sophisticated your employer's risk management monitoring systems have become.

The main types of hazards routinely encountered by the Digital YOU can be broken down into the following categories.

The Not Optimally Discreet Colleague

In our experience, relatively few people internalize the fact that you really can get into trouble for expressing thoughts on the job that were best left for a night out with the girls or boys. We often see people talking about illegal drug use, cheating on their spouses, having detailed sexual fantasies about their coworkers, and so on. While their behavior may be an open secret, a single lost lawsuit—even if it is in a totally different part of the company—can radically and without warning eliminate any tolerance for such things overnight. Further, quite a few companies sell extensive *libraries of terms* to trap even obscured "bad" content in text form. So, instead of George Carlin's seven dirty words, there may now be seven hundred that earn you a stern lecture if you use any of them in an electronic workplace communication. We have such libraries in a number of languages, including Klingon (yes, as in "Star Trek"). They make it easy to institute a difficult-to-circumvent policy quickly. And for those who are starting to get more clever about hiding their porn pictures, such as inside slide presentations, software like ours can actually extract the embedded images and compare their hashes[3] to a list of known porn images.

If you are wondering why investigators would care about this kind of stuff, the answer is threefold:

1. All things being equal, such behavior in the context of the workplace is generally indicative of broader judgment problems that may be of interest to us.
2. Sending such material to a specific set of coworkers is usually a sign of a high degree of social proximity or comfort, a dimension that it is often important to understand in investigations.
3. Several years ago, executives of a large corporate customer assured us that we would find "zero" pornography anywhere on their network because they had just had some very public firings for such behavior. You never want to use words like *zero* around us—it is tantamount to throwing down the gauntlet. Within about 20 minutes, we had found quite a bit of porn, but all of it was embedded within Microsoft Office documents of different kinds, making it much more technically challenging to detect. The text of the e-mails themselves was also clearly intended as camouflage; for example, "Hey Joe, I saw this presentation and thought you might be interested in it." It is, however, true that we found no individual pornographic images without the protective casing of a seemingly innocuous Office document. This is a simple but telling example of the general compliance arms race. That is, the deeply rooted contrary reaction many employees

have, when told that they can't do something in the workplace, is to try to figure out how to do that very thing without being caught.

Such subversion can easily become its own form of entertainment, quite apart from the porn itself or other forbidden activity. Unfortunately, when employers don't want you to do something, it's usually for a good reason—usually because they don't want to be sued. And they especially don't want to be sued and lose big. So it doesn't necessarily take a highly visible lost lawsuit for changes to happen. A sudden change in who heads the compliance, HR, or legal departments can cause guilt by association to be too much of a liability. We've even seen changes implemented because someone heard a nightmarish firsthand description of another company's eight-figure payout in a class action suit over a small number of dirty e-mails. Like many of the compliance-related hazards we will discuss later in the book, this one is dangerous because it probably doesn't matter at all—unless or until you get fired as a result.

The Compliance System

Highly creative rule-breakers are exactly the reason that compliance systems are on the lookout 24/7 for anything that could promote high blood pressure in the human resources department. However, most of these systems still aren't very smart. Many are just looking for the aforementioned seven dirty words, sexist language, racial slurs, and the like—in part because these are among the few things that almost everyone can agree are bad, regardless of the surrounding context. But these systems are going to be getting smarter—quickly—because, as you'll see through the course of this book, the motivations for them to do so are multiplying rapidly. The numbers of ways in which both individuals and corporations can be held liable (whether in a criminal or civil case) are increasing by the moment. This is because bad economic conditions almost inevitably bring more regulation; once put in place, such regulation is often there for good. However, from a purely psychological standpoint, what many of these general compliance systems are really for is to combat basic temptation—whether to wallow in self-pity, lash out in anger, bask in a cathartic moment or two, or engage in any number of other behaviors we all succumb to, perhaps frequently.

People I meet who are slightly clued in about how large companies are supposed to work are often surprised when they hear me say things like this. "Surely," they say, "most managers at these places are sent to compliance training and understand the danger of succumbing to temptation and putting anything over the line in writing." While that's true, it doesn't affect behavior much in the real world. It is really no different than the case of the normal guy who visits his doctor once a year and is told to eat less and exercise more—or else.

The first problem is the obvious one. Even if the doctor succeeds in scaring the patient, the fear soon wears off with the passage of time and the stresses of day-to-day life. Downing that triple ice cream sundae makes the patient feel better right now; it may or may not contribute to a heart attack 20 years from now when he might already be dead anyway. The second problem is a bit less obvious: the patient's sense of fatalism: "If I *do* die of a heart attack in 20 years, it will be because I have bad genes or they haven't found a cure for high cholesterol yet. Well, mostly."

The same logic applies to following through on compliance training. The rewards of a variety of bad behaviors are immediate, if small and short-lived; the fear is abstract and reinforced only infrequently. Further, a great majority of people seem to believe that the likelihood of their getting in trouble, regardless of what they do or don't do, is determined almost purely by whether or not someone important is out to get them—in other words, by politics, factors outside of their control. So there's no reason not to indulge in a little venting if it makes getting through the day a bit more bearable.

Unfortunately, all signs indicate that such indulgences are becoming more and more expensive, even dangerous. Not only can they land you in temporary hot water, but they can get you fired, sued, or even jailed.

You may be thinking that it is only human to sometimes be overtaken by small, mischievous urges—for example, poking some gentle fun at a blighted bureaucracy, an obsolete product line, a preposterously ugly design, or some other bag just asking to be punched. You have doubtless done it yourself, probably in the last week. Perhaps your motivation was the hope of bringing a smile to the face of a colleague who has been trampled by her ex-husband, two levels of management, and the market—all in the same day—or maybe you were just seeking a little badly needed commiseration with colleagues. After all, you've earned the right, haven't you? And all you did was say something that everyone else was already thinking and that was completely obvious anyway. It cheered up someone who really needed it—possibly you. What harm could it possibly do?

Ask Ralph Cioffi or Matthew Tannin. A couple of years ago, they were just two guys working on Wall Street in uncertain economic times, balancing their fears of failure with the need to stay focused and succeed in their jobs. However, what they now share the unfortunate distinction of being the first people ever arrested for contradicting themselves over a period of weeks in their electronic communications.

The Price of Contradiction: Cioffi and Tannin

Was a particular fund managed by the two Bear Stearns employees "performing as designed" or "toast" as part of the plummeting subprime market? Does the answer to such a question depend on the overall state of the market on any given day, the state of mind of the hedge fund manager, how many drinks he has consumed within the last hour, whether he is on his anti-anxiety medication, the target audience of the communication—or a combination of all these things? On June 19, 2008, the Feds decided that the determining factor was the target audience, since the negative commentary had been completely restricted to a trusted circle in Bear Stearns, and moved to arrest Cioffi and Tannin for fraud.

Yet the truth is they did nothing that most of us don't do, probably on a daily basis. In fact, if you're in sales, you probably do it at least a few times an hour. The men lamented the current state of the market; one wrote that he was "fearful of these markets," a statement that, given the prevailing climate, was an incredibly mundane thing to say—and doubtless a sentiment shared by virtually all of their clients. However, when talking with their clients, the crying in the coffee was unsurprisingly kept to a minimum, with statements such as, "We are very comfortable with exactly where we are." Not delighted. Not delirious. Not rapturous. Merely "comfortable," which could be considered a statement relative to what was going on in the rest of the world. Again, not exactly over the top—and certainly not what Court TV would call a smoking gun. But the combination of these statements, when coupled with a set of other unfortunate circumstances, was enough to completely disrupt the lives and careers of two otherwise completely unremarkable people. Not only that, it created a situation in which the line in the sand between a salesperson's optimism or spin and the legal notion of misrepresentation becomes very murky. Simply put, contradicting yourself in the wrong circumstances can not only be embarrassing; it can actually send you to jail.

Cioffi and Tannin were finally acquitted after a high-profile trial. Jurors interviewed after the trial effectively said that the evidence was contradictory and nuanced, that some days the mood of the two was bleaker, but on other days more cheerful. There's no law against being frequently moody or depressed when the market is going to hell. As defense lawyers trotted out more and more of the defendants' e-mails that didn't neatly conform to the simple, consistent pattern of premeditated deception the prosecutor was pushing, the case against them cratered.

The Cioffi/Tannin case will only be the first of many. The next prosecutor may have learned the key lesson from their acquittal: even a few apparently damning messages by themselves aren't enough to convict. Bad documents can always be shrouded with good (or at least ambiguous) context. A handful of documents paints a simple picture; dozens are likely to paint a more complex one. To really get the big picture, you need to master the full context in which various messages were sent.

Cataphora developed the visual analytic in Figure 2.6 a few years ago, well before the troubles at Bear Stearns. We already knew from experience that if you had enough data about a particular person from a long enough period of time, you would easily find a significant number of contradictory statements. While the possible reasons for the contradictions can range from the person having a good day versus a bad one, to an honest change in opinion based on new facts, to the simple need to say whatever is most expedient at the time, exposing such contradictions wherever possible is always a good way to rattle the other guy's witness or make him look bad. To help lawyers visualize this, we devised the chart in Figure 2.6, on the following page.

Each row represents an actor of interest. Each column represents a topic of interest whose presence in an e-mail or other document can be detected with a reasonable degree of accuracy by a computer program. If a particular actor generated no content on a particular topic during a specific time frame, the block where the actor and the topic intersect displays a diagonally oriented line. If some relevant content exists but is atonal—meaning it is bland, dry, and inexpressive—the block is solid-colored. However, if there is tonal content—meaning strong sentiments are expressed—

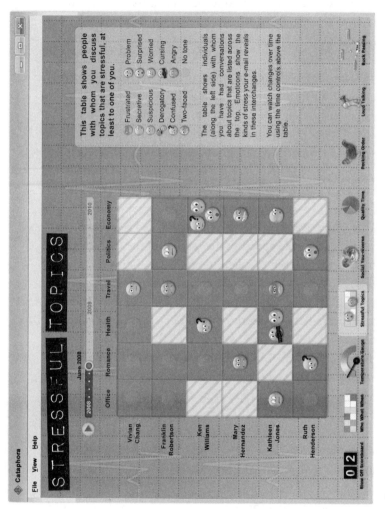

Figure 2.6 Stressful Topics matrix

the block is also decorated with the appropriate emoticons. Note that one of these emoticons represents reverse sentiments, or what we call "Two-faced"—in other words, the expression of contradictory sentiments that may be of interest to our clients.

Some of the other emoticons, such as Worried, Secretive, or Suspicious, are also likely to raise flags. Our software can detect the presence of such sentiments using a combination of techniques, including analyzing not just vocabulary but sentence structure, punctuation, and of course WHETHER SOMEONE IS USING ALL CAPS, because this is how Digital YOUs yell.

Back when we developed this analytic, we saw it largely as a clever piece of mischief in our arsenal. Now, however, we see it as a way of capturing the essence of the next generation of compliance problem: monitoring communications that, when assessed in the context of all relevant communications, could be considered fraud. It not only shows mood and contradictions clearly; it also depicts specific periods of time when a person was in an overall bad mood—or an overall good mood. By directly comparing different topics side by side, we can convey an appropriate sense of relativity. For example, this type of matrix would make it much easier to see that even if the traders were bearish on their own fund, their opinions on other comparable funds were vastly more pessimistic. It would also show who someone discussed their real frustrations with over time—as opposed to the odd off-the-cuff venting. All of this is information that would doubtless have been beneficial to the prosecutor in the Cioffi/Tannin case.

Clearly the two men must now wish they hadn't sent this or that particular e-mail—or probably any e-mail, ever, about anything other than the Yankees. But the real mistake they made was that their electronic communications essentially read like their diaries, reflecting their unfiltered and unvarnished states of mind as the market—and their employer—rolled and crashed around them. The fact remains that most of us do the same thing. Maybe we shouldn't have said this or that, but it was so easy, satisfying, presumably private, and harmless at the end of the day. It wasn't anything that everyone else wasn't already thinking, and it has to be better than liberating a whole box of Krispy Kremes. Most important, we think it will get lost in a sea of e-mails or instant messages, and besides, who would care anyway?

This type of case also exposes another subtle difference between the Digital YOU and the real you. In real life, for example, it is usually easy to see if you've been drinking, based on a number of physical cues. Unwise things you say when you are in this condition will generally be discounted. But you can be rosy-faced, slurring your words, and still thumb-typing on a BlackBerry in the bar on Friday night. Likewise, you may have giant bags under your eyes from many nights of insomnia because your fund declined in value by an amount equal to the annual budget of many small countries. The e-mails or IMs may not be noticeably less coherent than usual, but there's no Get Out of Jail Free card for the Digital YOU. Just about anything committed to bits—wherever you were and in whatever state—takes on a matter-of-fact air rather than an in-the-moment one.

The Banana-Peel Thrower

If you watch large-company politics play out long enough, it is hard not to notice that some people who are promoted up the chain take quite a few of their team members with them, while in other cases, there seem to be lots of mysterious career-stunting accidents among everyone who got anywhere near the newly promoted. The latter happens for a simple reason. Over the course of years in any work environment, a certain number of bad things will happen. And even if, somehow, no bad things happen, a certain number of mistakes will unavoidably be made in the routine course of business. In the face of bad things or mistakes, the usual reaction is to determine "who's going to get thrown under the bus," as one of our favorite clients loves to say.

A former boss of mine once generalized this phenomenon nicely. He characterized the banana-peel thrower as the most dangerous personality type in the corporation. His point was that whether someone throws a banana peel on the floor through sheer carelessness or because he is actively mischievous or malicious, if you fail to see the banana peel and slip on it, you are just as likely to break your neck. Being in close proximity to the banana-peel thrower is thus dangerous and to be avoided if at all possible. The place to be is next to the person who picks up the banana peels so others don't hurt themselves.

From our perspective, the person who picks up the banana peels is a guardian angel of sorts, a type who will likely inspire loyalty in those around her. Conversely, the banana-peel thrower may be so hated that witnesses may embellish or even lie in a deposition simply to make matters worse for him. However, the fact that he is a corporate climber or backstabber doesn't necessarily mean he committed wire fraud or corporate espionage.

The banana-peel thrower archetype has always been with us. But such people are now much more dangerous by dint of all the tools the wired world offers them: the ability to more widely project a presence, to leverage content so as to appear more knowledgeable than they actually are, and to mask their actual identity. This means the number of banana peels multiplies at a much faster rate than it once did. Luckily, all of these things also represent opportunities for such people to make mistakes, get in over their head, or get caught committing all kinds of passive-aggressive mischief.

The banana-peel thrower can most reliably be detected by an examination of changes in the social network around him over time. Social networks in organizational settings tend to significantly change for one of two reasons: changes in organizational structure or responsibility, and some kind of change in personal relationships. Otherwise, by definition, we have pretty much the same people performing pretty much the same tasks, so unless a personal relationship has gone up in flames, there is little reason for much to change over time. Further, even in the face of organizational change, social network structures can display remarkable permanence. A trusted source of advice is still a trusted source, even if she transfers to another department; groups of people who work well together will often reassemble if separated.

However, a personal relationship that has been truly soured will permanently change interaction patterns within a working group. The pattern created by such souring is readily identifiable by the fact that all the close links connecting the banana-peel thrower to other people weaken over time; they often stay strong just long enough for another unfortunate person to slip and fall on the next banana peel.

A graph can be used to visualize the types of "dead zones" that exist around habitual banana-peel throwers. In a time-based representation of the graph, links between two people are removed unless there is a significant and sustained amount of communication between them. Thus, in the case of banana-peel throwers who alienate those around them, the area around them in the graph is very thinned out, a bit like trees in a forest after a fire. For normal people, the longer they are at the company, the denser the social network around them will become. Alternatively, we can show snapshots of the work group at different intervals of time, so the movement of people away from or toward one another can be spotted easily. We can also highlight with a different color short-lived links, or situations in which communication ceased after a period of time even though both people were still in the same jobs.

This type of example highlights the importance in investigations of having data that spans a long enough period of time that repeated patterns of behavior can be observed, as well as having data from enough different people to have an accurate view of reality. For example, if someone were to take only relatively recent data (e-mail, IMs, etc.) from a banana-peel thrower and use it to determine who else to collect data from, he would overlook all of the people who had already been burned by that person in the past and were now keeping their distance.

The Corporate Echo Chamber

Some corporate cultures don't tolerate confrontation or dissent well (or at all). Generally, there are certain things you just don't do in most companies; what they are varies by company. If you do them, it's a sign you don't fit in or, even worse, don't care about fitting in. When I worked at a large company (which I won't name here), a very senior new arrival wrote a memo in which she harshly critiqued—demolished, really—months of work done by another group. It was abundantly clear that lots of time and money had been wasted and that all of the work would need to be redone. But what everyone was whispering about for weeks was not the work debacle, but the fact that the memo in question contained forbidden words like *incompetent*. The author of the memo was almost certainly shocked by how quickly and widely the memo spread through the company. In her mind, she was just clinically assessing reality, not poking a whole corporate culture in the eye.

This particular aspect of culture creates a powerful echo chamber that is invisible to the uninitiated. Note that sometimes even people who are well versed in a particular culture get caught in the echo chamber because they are missing some key piece of context. For example, imagine gloating over a rumor that a competitor is going to be sold off for small change to some foolhardy buyer. This surely seems harmless enough, at least until it becomes clear that the fool is, unbeknown to you, your employer. You may now get your proverbial 15 minutes of fame, but not the good kind.

Some of the content hitting the echo chamber is simply major announcements that are transmitted by various media with understandable rapidity. But more often, it is a matter of someone saying something she shouldn't have, whether it's repeating a hot rumor, sticking her virtual foot in her virtual mouth, or committing some violation of corporate etiquette. In pretty much any of these cases, from our point of view, it is worth knowing about. What an organization finds to be interesting or funny enough to circulate widely tells us a lot about both the character of that organization and the structure of the grapevine—for example, how closely it mirrors, or doesn't mirror, the organizational chart.

We visualize this dynamic in what we call the Flow of Information chart. This helps us see where such content originates, the path by which it is distributed, and whether or not it is generated by the same source and travels the same path consistently. In some variations, different topics of interest are color-coded, as some echo chambers are fairly topic specific in nature. Such real-world graphs are both large and complex; if you wish to see examples of them, please go to digitalmirrorsoftware.com.

Automated Employee Performance Analysis

Many people instantly react to the notion that a computer program, even a sophisticated one, could be used to assess their productivity. Obviously no one measure of productivity can capture everything. In many types of jobs, even determining how productivity *should* be measured is a fairly complex task.

However, computer programs have two advantages over humans in this regard. First, they are completely objective because they can neither like or nor dislike anyone. Second, they have the ability to perform large numbers of comparisons quickly—for example, to compare the amount of measurable work you have performed this year to what you did last year or to that of similarly tasked coworkers.

Jobs involving the creation, editing, or distribution of large amounts of content are the easiest to assess in an automated fashion (other than the simple tallying of some unit of work, such as a support call) for the following reasons:

- ◆ The amount of original content created can be measured; copied content can, in many cases, be identified.

- ◆ The number and kinds of changes to content can be measured. (Note that it is possible to roughly categorize changes to text; for example, fixing a typo or filling out a standard form are not the same thing as inserting substantive new paragraphs of text.)
- ◆ The frequency with which this content is actually seen, used, and reused by others can be measured.

Figure 2.7 shows a small example of how text is viewed in the Cataphora system; Figure 2.8 shows a somewhat more realistic view in terms of scale. Each rectangle represents what is called an N-gram, or a contiguous group of words. In this case, N = 2, so we have pairs of words. The lines connecting the different N-grams indicate the number of times one N-gram immediately precedes another in text. In this example, there is only one ordering of different N-grams,

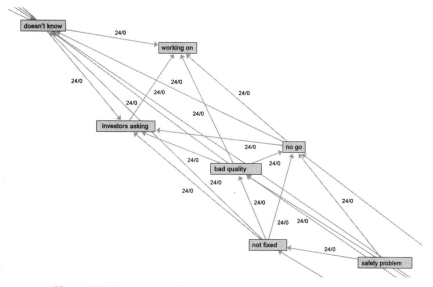

Figure 2.7 N-gram graph

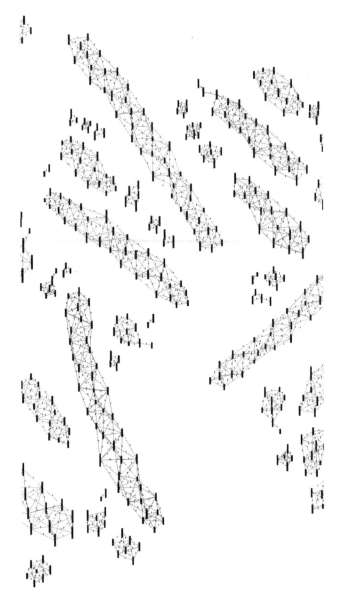

Figure 2.8 Zoomed-out N-gram view

which occurred in 24 different places. This means, for example, that "safety problem" precedes "not fixed" 24 times and that the reverse ordering occurs zero times. Using this network of N-grams, we can detect the first of these occurrences and identify the author, who is defined as the creator of the "original" content.

The key here is that productivity can't just be a measure of the quality of content created, because even if that content is good, if no one benefits from it, the exercise of creating it was not productive. Borderline cases can, of course, still occur. For example, it is possible that substandard content is being widely used. (If so, however, the real flaw is not one of personal productivity.)

An example of this is illustrated in Figure 2.9, which distinguishes between significant original content creators and *content curators*. Curators, in this context, are editors and distributors of content. They do things like customize presentations for specific customers, publicize the availability of new sales tools, and add that extra polish of proper formatting. Without the curators, lots of genuinely good content would never get into the hands of those who can benefit from it. Such people also often have a vast store of organizational knowledge, by virtue of which they can often help ensure that the right content is generated by the right people. There are usually far fewer high-quality original content creators in an organization than there are highly effective curators. As a result, much of the building-block content in many companies is written by a mere handful of content creators. The rarity of good content creators only makes the role of the curator that much more important.

Significant content creators are indicated by rings that appear to be black. The larger the ring, the greater the *quan-*

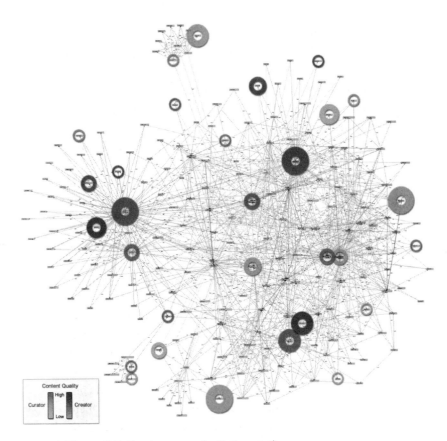

Figure 2.9 Employee productivity graph

tity of content produced; the deeper the black, the higher the perceived *quality* determined by the frequency of viewing, use, and adoption. Content curators are designated by gray rings.

So why do I list this type of analysis as a hazard? The answer is simple: any tool can be misused under the wrong circumstances, even with the best of intentions. For example, if we have several years of data for a given person, there will almost always be significant gaps in productivity. This is especially true of higher performers, since there is more room to fall off. The root cause is usually one of the obvious suspects: health problems (personal or family), divorce or other serious relationship problems, a death in the family, or financial troubles.

The reality in such cases is that the handling of longer-term performance issues is a management judgment, an inherently human task. As with the banana-peel thrower, this problem was hardly invented in the digital era. However, it has clearly been worsened by it. As already noted, the Digital YOU can't be noticeably drunk or visibly sick. Nor can it look like it hasn't slept in a week or had a month of bad hair days. The digital record captures most of what goes on, but not all of it. While the use of automated tools can make such productivity measurements more intrinsically fair, what they cannot hope to assess is the intrinsic value of a particular employee.

Getting Caught in the Web of an Investigation or Lawsuit

Few people ever consider the fact that if there's a lawsuit or investigation relating to the work done in their organization, a lot of employees' personal data is going to be collected—including their own. In addition, backups of data made over the course of months or even years will often be collected, removing the possibility of getting rid of things you subsequently thought better of saying. (We'll discuss the details of this scenario—and most important, what *not* to do if it should happen to you—at length in Chapter 7.)

A common misconception we encounter is that this type of investigation only happens to top executives. In fact, most of the cases we deal with end up involving data from hundreds of people for a number of reasons that are not immediately obvious. For one thing, if an adversary demands data as part of a legal proceeding, he is going to try to cast the net as widely as possible. This is a good way to maximize your risk, cost, and annoyance. Plus, the more data he can get, the greater the likelihood that he'll be able to find something to at least embarrass, if not actually incriminate, your side. Even more basic, lawyers rarely know exactly who within the "enemy" corporation to target, so it is in their best interest to try to haul in as much data as possible to be on the safe side. A judge ultimately determines what will be allowed and where the line will be drawn. Her objective is to require as much data as necessary to uncover the truth, but not more than that.

Nevertheless, to determine the truth, a massive data collection must often be done, because the real issue is the common business practices of the target company. In this event, data is likely to be collected from hundreds or even thousands of average people who were merely executing policy, sometimes as a means of establishing what the policy actually was. (Keep in mind that just because an executive says something, that doesn't necessarily make it true. What usually causes large damages—and hence lawsuits—is lots of normal rank-and-file employees executing policy made on high.) Or let's take a more perverse example: a class action lawsuit alleging discrimination against a broad group of people will likely cause data to be collected from all people belonging to that group to try to determine whether there is actually evidence of discrimination. But in the process of looking at all this data, it will likely be discovered that some of these people are among those who have buried porn in PowerPoint presentations or committed other termination-worthy offenses.

All in all, the digital world is getting to be a much scarier place. When I say that, people generally assume that I'm talking about stalkers preying on children. Most of them tend not to worry about all of their electronic actions being recorded, just as even savvy people sometimes put stupid things in writing: there is an assumption that unless you are a top executive, whatever you say or do will simply get lost in the infinite ocean of bits. This assumption is predicated on a fundamental untruth: no one cares about the little guy all that much (and probably not at all).

What is especially insidious about this untruth is that, for most of us, it is true most of the time—assuming that we avoid tripping the compliance system with all seven dirty words in the same e-mail. It is true until it is not true. The problem is you have no way to assess why someone will suddenly start to care about you or when. It is a bit like knowing that a cop may sometimes be hiding along the highway somewhere to catch speeders, but not knowing how often, when, or where. Of course, with sharp eyes, you may notice the cop hiding under that underpass and tap the brakes in time, whereas you simply can't see automated surveillance.

Some of you reading these words will almost certainly have had some of your data analyzed by our system at one time or another, whether it was because you were employed at a target corporation or were communicating with someone who was. And, really, you have absolutely no way of knowing.

3

Actions Speak Louder than Words

IN THIS CHAPTER, we'll discuss how the Digital YOU can reveal the real you's impressions about the world, particularly your opinion of others. The Digital YOU can get away with far more discriminatory behavior than you can in the real world. People can see their own interactions with you much more clearly than they can your interactions with others online, especially within the confines of a corporate environment. For example, people generally notice when you don't respond quickly to their messages, but unless they have some way of gauging how quickly you respond to comparable messages from others, they don't know how to interpret the delay. Nor do your friends likely have a complete list of all your user names or aliases on different sites.

But even if such widespread observation were possible, correct interpretation of large amounts of detailed data is impossible without fairly sophisticated technology. Consequently, the Digital YOU doesn't usually feel the need to be sensitive to others when such sensitivity is unlikely to be noticed by anyone. It is a bit like not putting on makeup or shaving before going somewhere where you don't expect anyone you know to see you; many of us don't exert the extra effort. We need someone else to see us to make the concept of not looking our best meaningful. As we'll see in the examples of digital "body language" that follow, having a Digital YOU that is optimally courteous and sensitive to a whole host of coworkers actually takes a tremendous amount of effort.

Important note: If you have not already done so, now would be a good time to go to digitalmirrorsoftware.com to download the Digital Mirror software. This will allow you to see how actions and words diverge in your own personal ecosystem—at least, the view of it afforded by your e-mail.

Twenty Digital YOU Revelations

1. How do you think about whom you include on your e-mails?

We rarely consider the order in which we enter data, but that order is often revealing. The first addressee is likely a very important and/or obvious recipient of the communication. The last addressee may be an afterthought. Since most people interact principally within the context of one or more work groups, it is possible to examine the consistency of data-entry order. We have found that the order of recipients doesn't change or is substantially similar over time.[1] Consider the following example:

To: Upwardly Mobile Joe
To: Stunningly Gorgeous Suzie
To: Stan the Cousin of the Big Boss

To: Bob Someone Must Actually Do the Work
To: Jason Grunt
To: Crystal Even Bigger Peon
To: Connie Intern for the Assistant's Assistant

This will often be the order in which e-mails to that particular work group are addressed (naturally omitting the name of the sender). To the extent that there is any variance, Jason Grunt is still unlikely to be the first-listed recipient; it is far more likely that Stunningly Gorgeous Suzie and Upwardly Mobile Joe will swap order, depending on the sender's priorities.

Everyone can see the order of recipients, but almost no one pays any attention to it. At best, people will check to see if they've been cc'd or appear on the To line, as that may determine how carefully they should read the message. Part of the reason recipient order is such an excellent measure of the pecking order is that no one feels any need to be politically correct about something that no one notices. This information is factored into the digital pecking order assessment in the Digital Mirror software. As you can see in Figure 3.1, the pecking order is shown from top to bottom. The position of the chicken designating each actor indicates just how much higher or lower he or she is from the next person in line.

The order of the speed-dial list on your cell phone—and arguably the extent of customization for each entry (such as a special ringtone or picture that indicates the identity of the caller)—is a non–e-mail example of this phenomenon, as is any kind of Friends list to which you have added people sequentially over time.

Figure 3.1 Digital Pecking Order assessment (*Grain art courtesy of Sage Ross—http://bit.ly/dqLScu*)

2. Save or delete?

We see examples of behavior ranging from obsessive e-mail packrat syndrome to those who keep their stored e-mail to the absolute spartan minimum. There is a variety of reasons for the width of the spectrum. In heavily regulated industries, some firms are compelled to save certain types of data for a very long time, whereas companies in extremely litigious industries encourage or even require the deletion of any nonessential scrap of virtual content. However, an increasing number of companies are now being advised by their attorneys that "less is more," meaning that the less information they keep, the lower their possible risks and legal costs. (We'll look at this issue more in Chapter 7.)

Therefore, many company IT departments impose space limits on the amount of e-mail users can store and then pelt an offender with escalating nastygrams—which, of course, further adds to the offender's inbox. Some companies even use software that marks messages for automatic deletion after a certain period of time unless the user specifically snatches a particular message out of its grasp. However, truly dedicated packrats will simply back up anything they want to keep to a local disk drive or somewhere else on the network, even if doing so is against the rules.[2] The lengths to which people will go to circumvent these obstacles demonstrates the value people place on their electronic communications.

While it is informative to look at the topics of e-mails that individual users choose to retain, such decisions are often largely motivated by specific job responsibilities rather than by personal preferences. In other words, there is a conscious thought process that kicks in and says, "I'm supposed to keep this." But if we focus on the identity of the senders instead—eliminating obvious junk e-mails that everyone deletes, as well as those that are generally preserved by all recipients—strong personal preferences for or against specific individuals quickly emerge. A comparison of saved e-mails versus those deleted shortly after receipt versus those removed subsequently when pressed by space limitations helps us determine which people's content an individual really values.

Most people delete unwelcome e-mails from colleagues they consider irritating and keep e-mails they are happy to receive. For example, one Cataphora engineer confessed that he had retained a lunch invitation from one girl for years, even though he had deleted many similar e-mails from other colleagues.

While it is easier to recognize this behavior through e-mails, it is not restricted to electronic messages. In fact, the attachments you take the trouble to save outside of your mailbox are also a good indicator of whose content you find to have some kind of value.

3. Whose feedback do you accept—or at least consider?

In many companies, etiquette prescribes the solicitation of feedback on draft versions of important documents. Invariably, some people's feedback is of far greater value than others'. Feedback that requires additional work will likely be ignored, unless the author either values the reviewer's opinion or simply thinks the suggestion is really good on its own merits. Simple and objective things like correcting typos are likely to be incorporated without much fuss. Aside from typos, whose feedback you choose to incorporate is a good measure of your professional assessment of that person. A computer program can attempt to assess this in a variety of ways:

- If someone suggests replacing "this purple polka-dot text" with "that orange squiggle text" in an e-mail, IM, or edited version of the document, the program can check to see which version of the phrase was present in the document the next time the author sent it out or checked it into a document repository system.
- If someone merely makes a comment about the purple polka-dot text but does not provide a replacement, the program may still look for any kind of change around the relevant text.
- If the author asks questions about the suggestion and/or solicits the opinion of additional people about it, this can reasonably be seen as evidence that she considered the suggestion.

Of course, some people may graciously thank everyone for the "great feedback" and never really use any of it. Others may feel an obsequious need to try to please everyone and will perform contortions trying to incorporate portions of everyone's feedback—even if some of it is mutually contradictory. However, most people fall somewhere in between. When they are short on time, they will prioritize feedback according to their perception of its importance.

Note that while some people will check later drafts to see if their changes were incorporated, few ever look to see if *other* people's suggestions were incorporated. Further, people rarely brag that their suggestions are consistently discarded, so the odds of getting caught are low. However, if technology is used to assess the objective impact of different employees in terms of content produced, such nuances will quickly be brought to light.

4. Whom do you frequently quote or otherwise acknowledge? Whose statements do you implicitly accept as fact?

When you explicitly quote someone, you are affirming that person's credibility and the credibility of the expressed fact or opinion, as well as affirming your excellent judgment in citing this particular individual. In a cynical scenario, you can use quotes to evade criticism if the statement is considered foolish or worse. Unsurprisingly, quote borrowers often tend to stay faithful to the same small number of sources, at least until something goes wrong.

Software can easily detect the copying and pasting of any amount of text that is long enough or distinctive enough to occur rarely. For example, the prior sentence may not be especially memorable or noteworthy, but the odds of exactly these words appearing in exactly this order without someone having explicitly copied them are still very low. Thus, if it turns up elsewhere, it has likely been borrowed.

5. Whose meeting or social invitations do you respond to right away and how?

This is another simple test that can expose the Digital YOU's lack of enthusiasm for certain events and/or people. Whether the invitation is social or work-related, if it relates to something pleasurable in some way, you generally respond quickly and in the affirmative once you receive it, barring an existing commitment. If the event in question is something horrifyingly dull and you can avoid it, you will likely send back a negative response with almost as much alacrity.

If you are lukewarm about the invitation, you will naturally take longer to decide on how to respond. You'll consider whether you have anything better to do at that time—

or whether, given a few days, you can manufacture some-thing—and whether you can get away with giving an excuse without offending the host. If there is already a conflict-ing obligation on your calendar and you still don't decline quickly, it suggests that you are at least musing about doing some schedule shuffling to accommodate the new invita-tion. The end result provides a good indicator of the relative value you place on one thing versus the other.

6. When and with whom do you use local vernacular?

Cataphora employs a number of Southerners. The more of them who are on the same e-mail thread, the more the Southernisms multiply. This usage indicates the level of familiarity among the e-mail recipients, not just with other Southerners, but with any poor Yankees or other "foreign-ers" on whom some of these colorful expressions might be lost. Further, the willingness of non-Southerners to borrow such vernacular is an indication of the speaker's influence. For example, at Cataphora, everyone uses the Southern expression "got their ox in a ditch." I attribute this to both the fact that it is a vivid and quaint phrase and that the par-ticular Southerner who introduced it, Jim Burton, is not only much-loved but joined the company quite early, which commands considerable status in a start-up. (The expression refers to someone who has screwed up in a manner that will take lots of effort to fix.)

Usage of local vernacular or dialect can do more than just cause some head-scratching on the part of outsiders. Such dialects identify the author as being from a certain region and perhaps also suggest his socioeconomic class and eth-nicity. This may help promote bonding among homogeneous groups of people, but it can cause isolation outside of that group. Black English is one example of this phenomenon in the United States, but other cases abound elsewhere. For example, in the German-speaking world, "correct," or region-neutral, German is usually considered *Hoch Deutsch*, or the German that is spoken in Berlin. All German speakers learn it, even if they use it only when speaking to those from regions or countries other than their own. By contrast, the German that is spoken just next door in Switzerland is headache-inducingly different. In the years following World War II when Germany was divided, a distinct East German dialect began to emerge. More generally, in some parts of the world, different regions have unpleasant histories with each other and tend to retain some lingering level of mutual distrust. Using your own dialect with people you don't know therefore potentially exposes you to the risk of being seen as an outsider, less advantaged, or less educated. Doing so in your personal or professional communications is a clear measure of comfort.

7. To whom do you respond when you aren't there?

In this increasingly wired world, people rarely truly fall off the grid. However, when you take off for two weeks of Tahitian bliss, the last thing in the world you want is contact with anyone you find tedious or annoying. On the other hand, you may very well respond to someone you think is delightful, even though you "aren't really there."

The people with whom you communicate when you are out of the office is an excellent indicator of who you like and, to a lesser extent, who you think is important. (Consider that your boss has no way to know definitively that you deliberately let your cell just ring and ring when he called. After all, perhaps your phone showed just one measly bar of reception in Tahiti.)

Similar logic applies to weekends and holidays, though it is usually harder to plead poor cell phone coverage. In both cases, we can compare how differently someone treats their colleagues—for example, responding to some on the weekend but waiting until Monday to respond to others.

When someone is out sick, especially with something relatively serious, they usually try to avoid potential aggravation. This makes them more likely to ignore the people they just don't like.

On a similar note, the people to whom you regularly respond from a mobile device, as opposed to waiting for the accommodating conditions of your office, is also an excellent measure of the importance you ascribe to them. This measurement can be further refined by considering only long messages and/or lengthy replies to messages, which are even more burdensome to read or respond to on a small device.

8. Who is important enough to make you break from a meeting or a train of thought? For whom do you almost always answer the phone? To whom do you wait to respond until it is more convenient?

In a world where it is increasingly common for people to show up at meetings and immediately open their laptops, it is also increasingly acceptable to try to interrupt someone via an instant message even if you know from his calendar that he is in a meeting. The question is whether the person allows himself to be interrupted or not, either shutting down his IM client for the duration of the meeting, ignoring the messages, or firing off a quick "Not now—I'm in a meeting" message. Similarly, if I am on a conference call at my desk or engaged in an active chat session with someone, do I try to squeeze you in or wait until I have finished with the other meeting or person?

A fairly accurate first assessment of this measure can be made by comparing online calendar entries with phone system records, chat session logs, and the time stamps on e-mails. In this fashion, we can see when you interrupted an activity such as a meeting or phone call to respond to someone else. We can also tally up these interruptions over time. In general, most people will find a way not to keep someone they care about waiting long and will pounce on any excuse to delay or avoid contact with someone whom they consider tedious.

9. How does the formality of your communications vary with different correspondents?

Formality has declined sharply in the workplace over the past decade. The reasons include the proliferation of both much more informal media such as Twitter and Facebook and devices such as iPhones, Treos, and BlackBerrys, which serve to discourage all but the most persistently and obsessively verbose from creating long messages. And shorter messages are almost always less formal than longer ones. Let's face it, formality takes significant effort and, like getting dressed for a black-tie event, requires exponentially more effort the less often you do it.

People who grew up in a formal business culture may be slow to adapt their habits to the tweet age and will continue to keep a formal tone in their business communications. But for everyone else, formality is limited to those situations in which they are trying to create or maintain some distance between themselves and the recipients. Simply put, the injection of an uncustomary level of formality is a means of telling someone that she has breached the Digital YOU's personal space.

My own communication style is informal, but every so often, when someone crosses the line, even though Cataphora is a relatively small and generally informal start-up, I will write an e-mail or a memo in full-on, big-company managementspeak. It works. The subject of my ire will avoid me for at least a week. (More important, the offender is unlikely to repeat whatever the bad act was anytime soon.)

Other times, the desire to put up the wall has nothing to do with the particular actions of an individual. For example, sometimes older people feel it is inappropriate for twenty-somethings to address them as peers and will try to find mechanisms to create what they believe is a proper amount of distance.

The use of formalism between specific individuals also provides an excellent longitudinal perspective on the levels of intimacy between individuals over time. People who work together closely abandon quite a bit of formality as they go along. On the other hand, if they are then separated and reconnect many months or even years later, the formality may creep back in.

Note that it is not difficult for software to detect various markers of formal speech such as "dear," "thank you," "kindest regards," "please feel free to contact me," "at your earliest convenience," and so on. In e-mail, it looks for whether there is any type of initial salutation at all; in a survey we did of data from a number of corporations, only about 40 percent of the e-mails began with a salutation. But superformal can be as revealing as plain rudeness of a strained relationship.

10. How much attention do you pay to spelling?

If you have to communicate in a language other than your native one, you are likely to make mistakes in grammar or spelling. If you are communicating in a language that has accented characters and you have an English-language keyboard, it becomes tempting to simply ignore inserting these characters. Unless you are a perfectionist or trying to impress the person with whom you are communicating, you won't spend the extra time and effort striving to be taken for a native speaker. On the other hand, if you sometimes achieve near perfection and other times hit Send with some errors or unaccented characters in your message, it provides a strong indication of how you view the different people in question.

The sad truth is that even native speakers often make spelling and grammatical errors in their given language. Software spelling and grammar checks can't completely fix this, but many people have become overly reliant on them. The problem is the software can't know, for example, whether you really meant to type *bang* rather than *bong*. Nor does it know whether a proper noun or an obscure word that isn't in the standard bundled dictionary has been misspelled. In this event, it also won't know whether the mystery word is singular or plural, causing grammar errors to be flagged incorrectly.

The interesting question is when people who tend to make these kinds of errors—whether in reality or in the sometimes astigmatic eyes of automated grammar checkers—bother to try to remedy them and when they don't. In our experience, people will often take greater care in writing to a customer or prospect than they will in communicating with a colleague or even their boss, perhaps because the boss already knows that—even after all these years—some elementary school grammar teacher is still in a state of despair over a particular employee. Within a peer group, noting when this type of extra care is applied can provide interesting data on who commands enough respect—whether for reasons of rank, influence, or popularity—that others take the time to impress them. (If someone never remedies these errors, then it is quite likely she is unaware she is making them in the first place, despite the visual cues provided by her software.)

11. With whom do you escalate or de-escalate the intimacy of communication media?

People who wouldn't shrink from sending a total stranger an e-mail that boldly implies at least mild acquaintanceship will often hesitate to pick up the phone and call someone they actually *do* know. In a sense, this is understandable. The asynchronous nature of e-mail lets you delay a response for minutes, hours, and possibly even days without looking foolish. Even with IMs, it is usually perfectly acceptable to wander off for coffee or otherwise disappear. But an actual phone conversation requires the ability to interact wittily in real time, a skill that usually atrophies quickly with disuse. Further, even if you are convinced of your own verbal eloquence, if you call someone who is less convinced of his own, you may risk provoking irritation. Another aspect of phone calls is that walking away can be difficult if the other party simply refuses to cooperate; most people dislike the idea of simply hanging up on someone they know even slightly, at least if they can't plausibly blame it on poor cell phone coverage.

What all this means is that if I want to minimize my contact with someone, I try to restrict the interactivity and intimacy of the dialogue to e-mail as much as possible. I respond to the communication when I have nothing more pressing to do, rather than at a time and in a medium of the other person's choosing. If he suggests a phone call, I politely demur, suggesting that e-mail is more convenient. If I think I will get more value from a greater degree of interaction, I may then try to initiate an IM session with him or even pick up the phone. This usually implies a sense of comfort, that there's no need to carefully craft every word in advance.

Software can track this in a number of ways, including using natural language processing techniques to understand when someone is either suggesting moving or refusing to move to a chat session or phone call. It can also determine via phone records and logs when communication occurred and who initiated it.

12. Do you leapfrog over communications from tedious coworkers to get to messages from people you enjoy?

Virtually no one reads or responds to e-mail in a completely sequential manner—at least, no one who receives any significant amount of it. The same is true for voice mails (with respect to responding), IMs, or any type of communications that stack up if left unattended for a few hours. Some messages may objectively be more urgent than others, but the reality is that most of us will jump to those messages that engender the most curiosity or pleasure, while deferring the more mundane and soul-numbing at least until after that first coffee or doughnut. And when you are about to leave the office or go offline at night, the temptation to simply ignore the dull task of responding to a coworker who is asking you the same question he asked you last week is almost overwhelming.

A simple way to assess this behavior is to examine the patterns of to whom you reply when you first get online, as well as just before you pack it in for the day. If we want to be more accurate than that, we can start to factor in things such as time differences, because there is less incentive to reply quickly if the person on the other end is likely to be asleep for the next eight hours anyway. Or we factor in the length of the response, since the need to provide a detailed answer may well affect the order in which you respond.

Although many companies encourage a high degree of awareness about how long it takes employees to respond to messages, there is rarely any sensitivity to—or even any glimmer of recognition of—the *relative* order in which they respond to messages from different people. Yet most of us brazenly express our personal preferences for some individuals over others by our choice of ordering.

The Blow Off Scoreboard report in the Digital Mirror software shows both how you treat others in this regard and how they seem to be treating you. The third column, as shown in Figure 3.2, will give you average response times.

13. Which people do you include together in what types of communication or otherwise put together in the same group?

It isn't terribly surprising that most of us correspond—using any media—with the same set of people over time, absent a change of job. But what you might find somewhat surprising is just how fixed the groups of people with whom we communicate are.

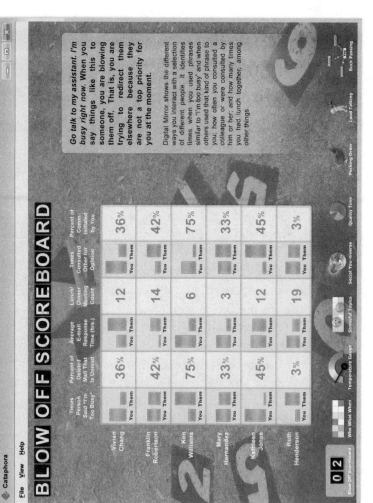

Figure 3.2 Blow Off Scoreboard report (*Background photo courtesy of Ed Sanders—* *http://bit.ly/abRmCl*)

To take a simple example, hobbyists within the workplace who have similar interests are likely to find each other quickly in these days of social media. So whether the topic is "Project Runway" or the NCAA tournament, if an enthusiast sends an e-mail, it will likely always go to the same set of people. Or a set of people that in most instances will only change slowly. As another example, if you are really upset about something your boss just did and are foolish enough to carp about it in an e-mail, you are likely to always send it to the same set of people. The same is true if you hear a really interesting rumor or are seeking advice about a potentially delicate situation.

By cross-correlating the set of people you send e-mails to (or invite into chat sessions) with the topic being discussed and the type of sentiments being expressed, it becomes easy to see what kinds of relationships you have with those around you. For example, many people have friends to whom they turn most often when they require consolation for one reason or another. And they may not mix the consolation friends with those eternally jovial types who never suspect that they need consolation.

You can use two of the reports in the Digital Mirror software to see whom you talk to about topics that are stressful to you, as well as with whom you talk about what. Figure 3.3 displays the most stressful topics and with whom you communicate; specific emoticons are used to distinguish between states such as Confused, Worried, or Suspicious. The degree of color saturation in the matrix in Figure 3.4 indicates how much communication you had on each topic with each group of people.

This kind of assessment provides a good, if simple, example of why it is so important during investigations for

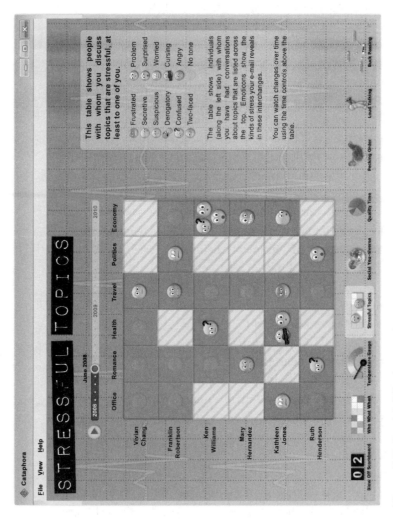

Figure 3.3 Stressful Topics matrix

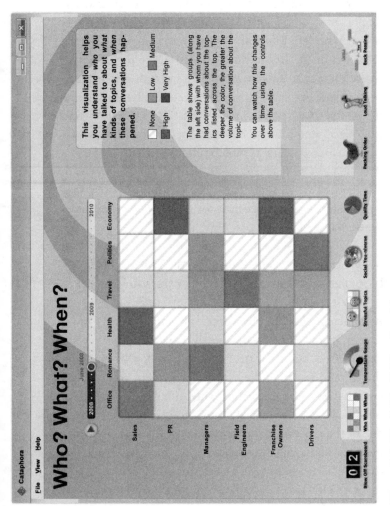

Figure 3.4 Who? What? When? matrix

multiple witnesses to be interviewed regarding the character of key individuals. Even if everyone interviewed is totally forthcoming and truthful, each person's perspective may be very different. While this has always been the case, it is unquestionably accentuated in the digital world, where there is at least the credible illusion of being able to keep separate groups of people separate. So the consolation seeker who may restrain her conversation in the real-world workplace—for example, out of fear of looking permanently pathetic to coworkers six months after a breakup—is far less likely to exercise that same restraint if all conversation is in bits that only a few individuals can see and hear (well, they and whatever compliance monitoring software is lurking in the digital shadows).

14. Whom can't you get enough of (Twitter feeds, blog alerts, and so on), and whom are you following closely enough to actually comment on?

As someone who typically works long hours, I have to admit that my gut reaction to any e-mail signature of the general form "follow me on Twitter" or "read my blog" is—nothing that should be committed to the permanent record. And even that pales in comparison to my enthusiasm for requests to join yet another social networking site. Nevertheless, one of the most insidious properties of the social network media is that saying no to an invitation can be awkward. For example, if you receive an e-mail from someone you know that explicitly asks you to please follow her on Twitter, what are you to do? After all, it takes only a few moments and is free. Strangers follow strangers. How can you say no and not feel the need to avoid the person in the hallway when you pass her?

So the more interesting measures of whom you can't get enough of involve actual investments of time, which is a precious commodity for most of us. If I am retweeting your tweets on an ongoing basis, I am ascribing a certain importance and value to you. The same is true if I regularly write on your Facebook wall or link to your content, and so on. If I trail off over time, that also means something when considered in relation to how I treat others.

While much of this information is publicly available, someone still has to care enough and have the tools to capture it. Otherwise, the most someone is likely to notice is your absolute level of interaction with him, but he will have little idea about the relative level. For example, if I largely disappear from your digital view for a while, you will likely assume that either I am very busy or that I am lying on a beach somewhere. In reality, I could have decided that you are more boring than the cooler new people I met recently.

15. Do you indulge in foreign talking?

When used by linguists, the term *foreign talking* refers not to people speaking in a foreign language, but rather to speaking—or writing—to people in a manner that suggests you doubt their ability to understand you. Foreign talking can involve things like the use of simpler words and grammar than you normally use, frequent repetition, and direct attempts to query comprehension (such as "Please read these instructions very carefully and let me know that you have understood them"). One way to think of foreign talking is as the exact opposite of what you would ever put into a message to your boss.

People will often use foreign talking with colleagues who seem to have difficulty following instructions, who seem to be on the slow side, or whom they do not respect. And, of course, colleagues who may not understand English too well. Rarely does anyone do this consciously. Rather it is a reflex that kicks in when there is a fear of information not being transmitted correctly. As such, it is a reliable indicator of someone's opinion of others.

16. Whose personal details do you keep around?

I am always surprised when a friend or colleague remembers my birthday or some other rarely discussed personal detail. What is really going on is that the person has cleverly stored the information in an application that, among other things, helpfully brings up the "Don't forget to wish Elizabeth a happy birthday" reminder. Still, whether the information in question was committed to an iPhone when it was heard in passing or gleaned from some public source, someone has to gather different bits of personal information over time, since they aren't typically all made available at once.

This behavior is commonplace in professions in which building personal relationships is important, such as sales. In other types of situations, the storing of such information indicates either a long-standing, valued relationship or the hope of using such information to build a bridge to someone with whom you'd like to have such a relationship.

Some people are packrats for this type of personal information and will avidly capture and store such data for everyone they know. However, if most of us bother to keep track of such information at all, it will only be for a select subset of people we know. Others will likely assume that we are just not the sort to remember anniversaries and birthdays. Until or unless they somehow stumble on our personal details data store and discover they are not on our A-list.

17. Do you attract loud talking (in e-mail and elsewhere)?

Most people don't read every e-mail carefully, especially longer, more detailed ones. Indeed, almost all of us know people who read e-mail while on the phone or waiting for one or more IM responses. What's the countermeasure to such attention deficit disorder? Using bold or underlined words, different colored text, all capital letters, and so on for the really important parts.

Now it's true that some people can't resist using such lovely techniques, even when sending messages to an elderly aunt who receives an average of three e-mails a week. But by looking at a broad cross section of someone's e-mails, it becomes easy to see whether such "loud talking" is associated with specific individuals, specific topics, or some combination of the two. Note that loud talking usually isn't about questioning someone's competence so much as wondering whether she is actually paying any attention to you.

You can use the matrix in Figure 3.5 to see how often you use loud talking with other people and also who uses it with you. You can also compare how much loud talking you do at someone versus what everyone else seems to do to him—at least from what can be seen in your own e-mails. (Of course, this would be much more accurate if you had the e-mails of everyone else involved.) If we're doing an investigation and someone "loud talks" at you a lot, we can note whether or not you appear to be the exception or the rule. We'll discuss loud talking in greater detail in Chapter 4.

18. Do you feel the need for a digital face-lift?

Fashion magazines constantly exhort the view that a new pair of shoes or a trip to the salon is the fastest and surest

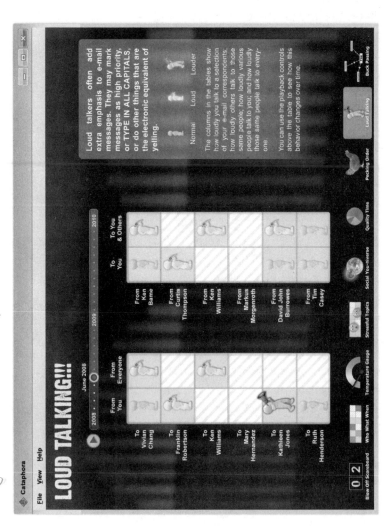

Figure 3.5 Loud Talking matrix

way to wipe away the blues. I am told that Coco Chanel once opined that there were "no ugly women, only lazy ones." In other words, with enough effort, anyone can be at least somewhat attractive.

This begs the question of whether enough attention—or perhaps money—spent on your digital persona can keep you from appearing dull or boring. Or washed up. Or like everyone else. When is it that most of us feel the need for a digital makeover? The possible or actual loss of a job, a failed relationship, or some other intense life experience are usually the situations that prompt a desire to overhaul your LinkedIn profile, Facebook page, and so on. Plus, you can always get a totally fresh start on whatever new community sites have recently turned up, thereby avoiding having a potentially unpleasant list of recent changes.

Many people are constantly refining their digital image. However, that is not the same thing as trying to rebrand yourself—for example, changing how you fundamentally describe yourself on a dating site or in a career profile. Nor is it the same as launching a shameless campaign for recommendations on LinkedIn or similar sites, which in my experience people tend to do either when they are feeling insecure, are thinking about looking for another job, or both. (As an employer, I can tell you that such recommendations carry negative weight. As a practical matter, most people who are asked to will dash off a recommendation as the path of least resistance, regardless of what they actually think. Further, many such recommendations are mutual: I warmly recommend you; you return the favor. What the employer sees is a mutual admiration society, not a meaningful recommendation.)

I am always surprised by how many people fail to understand that their friends and employers will readily pick up on such beautifying tactics. Unlike many of the other behaviors on this list that no one ever seems to notice, everybody seems to notice this one. The increasing number of feeds or alerts of different types all but ensures you'll be found out.

If you are thinking that this is absolutely no different than coming to work with a new outfit and a completely different haircut, by which you are clearly signaling a desire to change something, you're partially right. There is, however, one important difference. The haircut will soon be forgotten—unless you're a 60-year-old with a Mohawk—but the burst of online activity and revisionism will likely be there forever for prospective future employers, business partners, and romantic interests to peruse.

19. Whose pictures show up the most in your content?

Uploading pictures and video clips to many social networking sites is easy to do. Most of these media are "tagged" by the user, meaning the people in it are identified for easy search and reference. For people who do this fairly continuously, the number of tags that show up for each of their friends provides a decent barometer of who they are spending their time with during a given period. People generally post pictures online to show their friends what they are doing, to feel part of a community, and to make it appear that they are having a rip-roaring good time. It is unlikely that anyone would go from page to page and site to site manually counting tags. However, for a computer program that can access the various sites in question, it is easy to keep score.

20. What is the sheer volume of actual responses in words, minutes, and so on?

Unfortunately, most e-mail programs don't make it easy to see how long someone actually spent writing an e-mail, and it is the same for instant messages. This is a shame from our point of view, since seeing how many times words were erased and the message begun again would be an interesting metric, as would how long someone simply stared blankly at a big, empty white space while trying to figure out what to say or how to say it. We have to settle for indirect measures such as the volume of text content that someone sends to various colleagues and friends. From this, we can also try to estimate time spent by assuming an average typing speed. In the case of instant messages, we can make some inferences to try to assess the number of minutes actually spent responding.

For example, if someone is really engaged in a chat session, the mean time between responses will vary by only a limited amount, especially if you factor in the number of words in the message. Much longer lags are more properly understood as absence. Combining these with phone records and calendar events that indicate face-to-face time spent with a particular individual allows us to reconstruct with moderate accuracy how much time someone spends with different people. Since time is a limited commodity, this is perhaps the best indication of social proximity. Certainly there are people we could do without but with whom we must interact professionally, but usually the difference in time spent with them versus people we actually like is immense.

You can use the Quality Time chart in the Digital Mirror software, illustrated in Figure 3.6, to see how you are div-

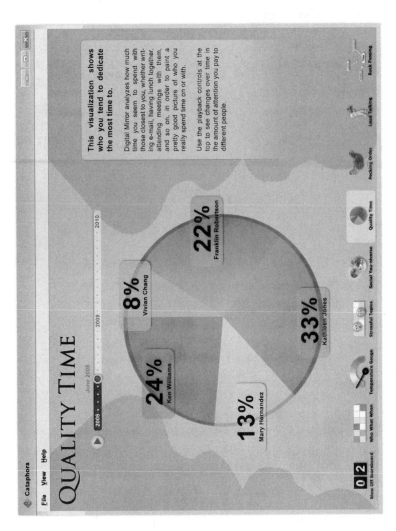

Figure 3.6 Quality Time chart

vying up your quality time. This information is ascertained from analyzing your Outlook data (which includes not just e-mails, but also meeting invitations).

If you are wondering why we would care about this in an investigative context, the answer is simple. How well liked someone is understandably has a significant impact on both how well they like their colleagues and their employers. For example, people who are generally well liked are less likely to be whistle-blowers or to leak confidential information out of malice. They are even less likely to be hostile or uncooperative witnesses in a legal proceeding.

4

Annoying Digital YOU Character Traits

IN THE REAL world, as in the digital one, we are often slow to recognize our own annoying habits or personal choices that may not paint us in the best possible light. Witnessing someone else commit an even more grievous version of that same error may jolt us into awareness. However, nothing works better than overhearing others' negative observations—especially memorable or witty ones—about the behavior in question. Here are some common real-world examples:

- People who talk endlessly about their children in response to a pleasantry such as "How's the family?"—or, indeed, with no prompting at all
- People who make dubious fashion choices, such as that Hawaiian shirt with the neon flowers or the orange hot pants (or the beard that prompts repeated references to the Unabomber) that may not make the anticipated bold fashion statement
- People who complain about virtually everything without any touch of humor
- People who yell, scream, or otherwise become visibly agitated at the slightest provocation

Once such habits or choices become the butt of common jokes, people can become increasingly sensitive about them. In the real world, the number of people who can listen to endless complaining or tolerate horrible clothing is limited. The bad news, of course, is that the digital world imposes no such limitations in either time or space. An ill-considered rant can linger indefinitely. A ludicrous fashion choice could end up featured on some worst dressed, most pathetic slobs on the planet website and be seen by a huge number of people, even if only for a short while. Worse, direct feedback from the surrounding universe is often hard to come by in the digital world; at least it's not as direct or immediate as it would be in the real world. An e-mail can be quietly ignored or punted to someone else. If you find a blog boring or offensive, you need never return. Digital feedback does exist, but most of it is indirect. In this chapter, I will try to paint a picture of some of the digital-world personalities you don't want to be.

Just as there are clichés of different kinds of clownish social behavior, especially in large corporate environments (see the 1999 movie *Office Space*), the digital world is unfortunately rife with new archetypes. Some of these are just direct projections of their real-world equivalents, exhibiting bad behavior that may be encouraged by various aspects of the online world (such as anonymity); others have no workable real-world analogue. But one thing is certain: being "raised well" doesn't necessarily translate to having good online social skills.

From an early age, most of us are taught that certain behaviors are unquestionably impolite and should be avoided. But judging by the evidence, few of us stop to ponder which of these behaviors also apply in the digital world. For example, it is generally considered rude to correct a person's grammar or pronunciation during spoken conversation, especially if that person is not a native speaker of the language. Yet you will see many people in online exchanges correct similar errors. Making fun of someone (other than a friend who will take it in stride) is likewise something that most people avoid in the real world. But relatively few restrain themselves from posting mocking comments in response to a blog or other online content with which they disagree.

Nor does sensitivity in the real world necessarily imply much about the digital one. A major reason for this is that the reactions of others are often delayed, and key cues like facial expression or vocal intonation are usually unavailable. Consider, for example, that in the physical world, we have many indications that someone has taken particular care with her appearance on a given day: every hair in place, nicer-than-usual clothes ironed and pressed, shoes shined, color-coordinated socks, and so on. In such a circumstance, you may be tempted to tell the person she looks particularly nice; at worst, you say nothing.

But in the digital world, if someone provides some new content on one of her personal Web pages, it is impossible to know how much care or effort she put into that content, making it seem less dangerous, and hence easier, to criticize. For example, the home page of someone purporting to possess expertise in setting up websites for small businesses might be quite slick visually but have sloppily written content. This may simply mean that the page is still under construction or that its owner had a friend take care of the graphics while the owner continued to work on content that is comparable to the visual aspect. Or perhaps it means that the owner should stick to his day job.

This is not to suggest that the Internet turns sensitive people into social cretins, but rather that certain types of misbehavior are simply far easier for the Digital YOU to commit—and to a far greater degree—than the real you, whether you know it or not.

The mini-portraits in the rest of this chapter describe the most common Digital YOU archetypes we've run across in our work. We have become familiar with these archetypes because, when investigating real-world events through electronic data records, we do many of the same things real-world detectives do in their investigations. Specifically, we use our whole battery of tools and techniques to help us truly understand the character of the key individuals involved. As I asserted in Chapter 2, character is destiny: it constrains the likely—or even possible—set of motivations and actions of each individual. A coward is unlikely to behave bravely or perform a feat requiring great nerve, a hot-tempered person is unlikely to behave calmly in the face of a severe crisis, and so on.

These archetypes exist everywhere and in the most normal and mundane circumstances. We all know them and live with them. Indeed, some days we *are* them. So let's take a look at some of the most common of these personalities.

The Digital Buck Passer

This is the person who, upon receipt of any communication, is congenitally incapable of making a decision and uses the magic e-mail Forward button or other means available to foist the decision off on someone else.

The telltale sign of this archetype is large numbers of forwarded e-mails with a minimum of added text or the equivalent behavior with other media; for example, IMing a link to someone and asking what she thinks the appropriate response is. Added text will be brief and sparse in meaningful content, containing nothing remotely resembling actionable advice or even a vague opinion. (A variant of this is an unreasonably large amount of added text that nevertheless lacks any actual instructions or substance.) In the case of e-mail, slick perpetrators will disguise this lack of substance with personal flattery. For example:

> I just saw this great opportunity come up and knew you were exactly the right person to run with the ball. What do you think?

or,

> This looks like it could be really interesting, but I'd like to hear your thoughts on it.

This may sound benign enough, and if the same person hasn't sent a host of similar messages, it may be. However, where we find even a small number of these types of messages from a particular individual, we rarely find any evidence of him making any kind of decision. This reluctance is not an attempt to be shady. Rather, many people in the corporate world are afraid to admit when they don't know the answer to something. And even if they think they do know the answer, they are terrified of having to take responsibility for a decision that could later turn bad in any of a thousand ways.

Virtually all of us have, at least once, hastily forwarded an e-mail or posted a request for help to some forum on our way out of the office on a Friday night, meriting a "What the hell do you want me to do with this?" response. Hopefully, though, that's the exception. But that Forward or Post button can become seductively easy to cozy up to, especially if you are a manager. Almost no effort is required, and often, as if by magic, the desired results will be produced. It is only much later that the buck passer realizes he has become little more than an information routing system—a position that is easy to dispense with when that next big layoff happens.

The instant gratification provided by the quick fix of passing the buck is addictive. In chronic cases, one-third or more of a person's communications are buck passing. The addictiveness of this behavior means that just one or two of these messages are often symptomatic of a broader pattern of evading decision making, or at least of recording decisions.

One interesting thing about this type of behavior is that it generally occurs completely outside the formally defined organizational structure. For example, a boss may not want to admit to an employee that he has no idea how to handle a particular problem. He may try to approach someone in a more senior position, ostensibly seeking mentorship or a bonding opportunity, when the real objective is to get a suggestion on the best way to proceed. In the case of two peers, one may couch a request for advice as promoting a spirit of collaboration and being a team player.

If the buck passer is successful in getting someone else to assume responsibility for performing a task or getting an answer to a particular question, he will often simply usurp the answer or the results of the effort, without crediting the real source. This is more likely to happen in groups of people who are geographically distant or in different parts of the organization, since the risk of the stolen credit being detected is much smaller.

In another scenario, the buck passer repeatedly thanks the person who actually did the work. There can be various motivations for such public acknowledgment. The buck passer may be genuinely grateful for the other person's contribution, but this is also someone to whom he can deflect any follow-up questions. And, of course, there is the option of passing the blame and saying, "It wasn't me," should the need ever arise. A more timid or meticulous buck passer may not settle for an answer from just one person; he may require—and publicly acknowledge—as many as he can find.

The prototypical "hands-off" manager regularly passes the buck to his employees. Yes, managers are supposed to delegate to their employees. But there is a big difference between delegating the responsibility for ordering the sandwiches for a lunchtime meeting and farming out the selection of a vendor for a $2 million contract. When delegating, it is a manager's job to provide guidance on the important decisions and tasks. Providing information such as traps to avoid, limits not to exceed, known best practices for approaching a problem, or other specifics is passing the baton rather than the buck.

Buck passing is so commonplace that the corporate ecosystem has adapted to reflect it, with the digital buck passers forming symbiotic bonds with individuals we call *power accumulators*. Power accumulators are the people who, for whatever reasons, are unafraid to make many of the day-to-day decisions that buck passers lack the courage or conviction to make. Often, this correlates to their actual title or job function, but not in the way you might expect. Those cloaked in the corporate anonymity of a lower-level job function, and hence unlikely to attract much notice, often find decision making to be less stressful than do their more visible and better-compensated superiors. But a real danger lurks here; the implicitly anointed decision maker may lack a critical piece of context or expertise that is needed to evaluate a situation properly. Nor is the power accumulator's compensation necessarily impacted by choosing either wisely or poorly. This is a real danger to the company, as it can result in even important decisions being made by someone who lacks the key expertise needed to do so wisely. It is also a danger to the buck passer and—perhaps most of all—to the de facto decision maker. Not having been noticed much before something bad happened doesn't mean you can't be fired for it afterward. In fact, it is usually easier to fire someone who was generally under the radar and has less managerial responsibility.

Many types of business decisions require input from different functions, such as legal, accounting, and human resources. Input from colleagues in different countries may also be needed. Difficulties often arise when a junior person in one of these functions needs to obtain an opinion from another function and simply doesn't realize it. For example, a junior finance department employee might say that a particular accounting treatment is fine, because it is correct in the United States, not realizing that she also needs to consult her colleagues in Germany and Japan. Or she might not realize that, for this particular type of transaction, the legal department must be consulted before she can sign off. But once "Finance has approved it," the train proceeds on its course and—possibly at some point—over a cliff.

While the most common reason for this type of situation is that the buck passer simply didn't want to be bothered with the tedious task of chasing down arcane details of international tax law, it can also be engineered by someone with a nefarious mission. A more senior person in the finance department who regularly dealt with international issues would be well informed when there was a need to verify the proper procedure in another country. So he could not credibly make the mistake of not knowing that such verification was needed. But what he *can* credibly do is pass the buck to someone more junior without that all-important instruction, thereby all but ensuring that the wrong thing will happen.

That said, most types of buck passing are the result of someone trying to avoid making a decision that could later lead to unpleasantness. Sometimes, the potential unpleasantness is small, if it comes about at all, but to someone who is inherently uncomfortable with making decisions, even a small amount of unpleasantness can be too much. Choices of carpet and wallpaper color in a new office are a good example. Personal tastes in aesthetics vary considerably, and the results of the decision will be around for years for at least some percentage of the occupants to dislike.

Laziness and the desire to avoid a tedious task can also be a motivation. However, buck passing is usually not about laziness, since often the laziest thing to do is to make a decision quickly and carelessly. Some types of buck passing are highly contextual in nature. For example, if a long e-mail with attached documents shows up on your BlackBerry or Treo, the temptation to transition it from your inbox to anywhere else can seem overwhelming.

One time when even non–buck passers often rush to the Forward button is when they receive a communication from a referred job seeker. (This is especially true when the referrer occupies a high-status position within the company.) The recipient of the résumé may feel the need to fulfill the obligation and send the résumé somewhere, but he is unlikely to cash in any favors to help the candidate move forward.

Figure 4.1 depicts the myriad communications that involved decisions in one of our real-world cases. The power accumulators are easily recognized by the gray spots, which are actually clouds of arrow tips pointing to them from digital buck passers. As is always the case in our experience, the number of power accumulators is small relative to the general population, and one power accumulator is often tightly linked to another. Simply put, the same individual acts as both a power accumulator and an intermediate buck passer. It is also common that few of the power accumulators' roles in reality correspond to their positions in an organizational chart. However, an individual with the appropriate managerial title and portfolio is never far from a power accumulator in a social network sense, whenever the manager is not himself a power accumulator.

The *hidden organization effect*, whereby information and decisions flow according to well-developed ad hoc channels, is usually pronounced enough that we use two organizational charts in most investigations. One is supplied by the company we are examining and reflects its formal structure; the second is generated empirically by our system. Our chart is based on the *real* decision makers, the usurpers, and the buck passers or anyone who was otherwise excluded from decision making, regardless of their formal titles. You can decide which of these is the "real" organizational chart. What is truly interesting about this duality is the possibility that the official chart is largely accurate with respect to understanding visible, real-world events, such as who gets invited to which meetings, but is surprisingly inaccurate as a tool for understanding what is true in the digital world.

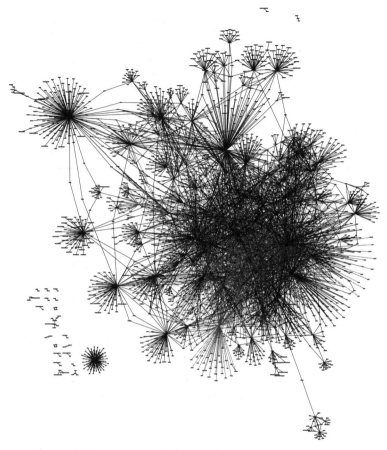

Figure 4.1 Power accumulator graph

Digital buck passing can be thought of as a careless and lackadaisical form of delegation in which no guidance is offered. The real you would probably never execute the drive-by maneuver of tossing a large file folder on a colleague's desk and saying, "You go figure it out," and then quickly departing. For your digital self, the world is a far more nuanced place. That link in the document *might* point to 500 pages of online content that need to be read and absorbed, or it might be five paragraphs explaining how the sky will fall in on you if you make the wrong decision. You won't know the difference if you unload it quickly enough on someone else. Nevertheless, in both the real world and the digital one, habitual buck passing is the antithesis of leadership.

That said, the relationship between buck passer and power accumulator is often a truly symbiotic one. The buck passer profits from the other person's decision-making ability and/ or other competencies that he wouldn't otherwise have—all silently transmitted straight to his desktop. The power accumulator gets the advantage of more control without many of the attendant stresses. True, the power accumulator probably doesn't get the big bucks, but she is also less likely than the usually far more expensive buck passer to be out on the street if the company has a couple of bad quarters. And she gets something else: the awed respect from her peers who wonder how she managed to invert the workings of the corporate food chain. Nevertheless, as in any symbiotic relationship, if you really *require* the other person to be present and performing her function, you are ceding your independence in a way that you may regret.

Tip: You can use the Digital Mirror software to see just how often you pass the digital buck—or are the recipient of buck passing. As shown in Figure 4.2, the arrows indicate the direction in which the buck is being passed between two people, while the respective sizes of the arrows indicate the relative amount of buck passing.

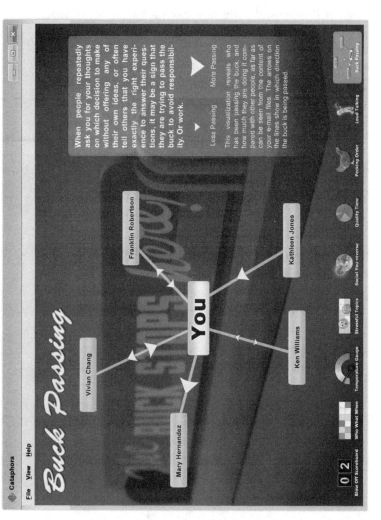

Figure 4.2 Buck Passing view

The Love Bomber and the Digital Stalker

The love bomber is the friend or colleague who shadows you online, responding instantly and often superfluously to your e-mails. She seems to always IM you the moment you show up online, posts comments on content you post, always sends you invitations to events, responds affirmatively to others' invitations right after you do, and so on. For such people, more communication channels and forums only mean more opportunities to reach out and smother you.

Love bombers, especially if they are your colleagues, may truly believe they are being hyper-responsive, indicating their profound eagerness to help out at any juncture. Sales-people may vigorously pursue a hot prospect online nearly to the point of stalking, a level from which they would shrink in the real world. And then, there are the old friends who feel you are abandoning them, not to mention the person who demonstrates that there are exceptions to the rule that "what happens in Vegas stays in Vegas." The difficulty here is that, just as there is recognition in the physical world of an adequate physical distance between two people and not violating someone's personal space, the same construct logically exists online. What is missing are generally accepted social conventions to govern behavior.

Consider that there is often social pressure to include everyone in your work group or bowling league in your Friends lists on Skype and other online contact lists. Although you can usually set privacy levels so that certain people cannot see when you come online, it can be an imposition to have to figure out how to block a particular individual. Likewise, many companies use fairly open online calendars to facilitate scheduling, thus allowing a broad cross section of people to see where you'll be, at least some of the time. Of course, in a professional setting, anyone can easily observe at approximately what time you normally start responding to e-mail and phone calls, as well as when you knock off for the day. And a great variety of Internet sources can provide data as to who has previously attended—or is slated to attend—specific conferences, charity galas, and countless other types of events. Not to mention the fact that many people frequently provide updates on their doings in various forums, such as Facebook and Twitter.

Some of the telltale signs of this archetype involve significant asymmetry in their communication patterns with targeted individuals: they reach out far more than they are reached out to; they save every shred of communication from the target so they can leverage it again later; and they continue to pursue their target even when they don't receive any response. Take the following IMs received from one person on the same day as an example:

> 1:38 [Target recipient shows up as "available" or online.]
> 1:42 Are you there?
> 3:09 Where are you? You weren't in that meeting.
> 5:17 What about going to the game tonight? Wait—I'll try you on your cell.
> 8:07 Now you wouldn't be standing me up for that blonde would you? ;-)

The more ruthlessly all channels—e-mail, cell phone, work phone, fax, message with the secretary, and so on— are used, the creepier and more irritating it becomes.

Whether the motivation is emotional—a craving for affection or just attention—or purely professional, a needy Digital YOU is just as unattractive as its real-world equivalent. But the consequences of a desperate Digital YOU often exceed what is likely, or even possible, in the real world. Consider the difference between the vague sensation that a lonely coworker is watching you with a bit too much interest from across the cafeteria and being able to fill a ream of paper with all the e-mails and IMs she sends you in the course of a month, most of which would be tough to characterize as solely work-related.

Most important, an employer is far likelier to take some kind of action in the latter case if you express any discomfort. It is disturbingly easy to overstep the boundary in the digital world, especially since the amount of actual effort expended by the pursuer may have little correlation with how pursued you feel. And indeed, the more tools there are to facilitate pursuit, the more the asymmetry of experience grows. Hence, the increasing prevalence of the digital stalker, especially in sales-related roles.

Here's one real-world example, with the names changed to protect the perps:

> Good morning Elizabeth,
>
> I'm sorry that I missed you again today. As I mentioned, Joe Reeves, my director, asked me to contact you and set up a meeting with you in the next few weeks. He understands that you have an extremely busy travel schedule at the moment and was, therefore, hoping for time with you in May or June.
>
> Joe also asked me to suggest a few dates and times that he is available and has asked that you please do the same. Joe has time on May 16 or May 28. In June, I have many times open. His best days to meet are Wednesdays, Thursdays, and Fridays. Should you have

any further questions or need a different date and time, you may reach me by phone or e-mail. I look forward to speaking with you soon.

Cheers,

Joanne Hardley

Flippowidgets, Inc.

This is a perfectly normal-seeming business communication. What makes it an instance of love bombing?

- I (the target) had never previously heard of Joanne, Joe, or Flippowidgets.
- This e-mail was part of a repetitive multichannel attack, involving a combination of multiple phone calls, voice mails, and e-mails. Each successive e-mail changed the dates a bit.
- The e-mail makes an implicit reference to knowledge of my schedule and to "any further questions," suggesting at least some prior contact, which did not exist.
- It is bold insofar as it did not try to deceive a secretary or receptionist that these people know me, but hoped to fool me directly.

All in all, it doesn't exactly make you want to go out and buy a flippowidget.

Here is another, more subtle digital stalker example:

Elizabeth, I'm sorry that we were unable to connect last February, but I would still love to have the opportunity to introduce Glibglab.com to Cataphora. Are you available to chat sometime this week or next? If so, please let me know a time that works for you, and I will arrange my schedule accordingly.

I look forward to connecting,

Kate

What made this one particularly clever and insidious was that the message contained a moderately long e-mail thread, again with the aim of implying prior interaction. However, on review, this interaction consisted of three completely unsolicited e-mails that Kate had sent me at different times. Since most busy people I know really hate to scroll, I'll bet this technique often works.

The most creative digital stalking attempt I've seen to date was an enterprising salesman who correctly guessed the e-mail address of our internal sales alias. He sent us the following message:

Hi All,

Marci and I just got back from this terrific vacation in Costa Rica. If you've never thought about visiting Costa Rica, I just can't tell you what you're missing: beautiful beaches, friendly people, and great seafood. We hope to return there the next chance we get.

I hope you all have been well, and I'm truly sorry that I let so much time go without scheduling a meeting with you to discuss how we might best partner together. Please give me a call when you can.

Regards,

Charles

P.S. I attach some pictures that Marci took from our hotel balcony so you can see how beautiful Costa Rica is.

Each of us assumed that someone else at Cataphora must know this Charles person. But as it turned out, none of us did. The approach, however, was an ingenious one.

The Guy Who Wants to Take Everything Offline

Many people still crave the intimacy of face-to-face conversations. Whether motivated by a sense of loneliness or isolation, or the desire to maximize the number of available physical cues in a business situation, they often try to bend your work habits and preferences to suit theirs.

While in some ways related to the love bomber, insofar as he craves contact, this archetype can perhaps best be thought of as someone who is not fully at ease yet with his digital self, much less anyone else's. The huge obstacle the "offliner" faces is that in a world where anything that requires so much as a walk down a corridor may be seen as inefficient, his desire for doing things the old-fashioned way may be difficult for others to accommodate. While a phone call should be a happy compromise, the really hard-core Digital YOUs may avoid this because it limits their ability to multitask. For example, many people can keep several unrelated chat sessions going while answering the odd e-mail in between. Talking on the phone renders this next to impossible—plus the offliner can probably hear the other person clacking away on those keys.

Such differences in communication are often attributed to generation gaps, but there are, in fact, many reasons for them. Some are vocational in nature. To take an extreme example, a police officer would never choose to interrogate a witness online—far too many critical visual cues would be lost, and an evasive witness would have more time to craft his answers. Doctors will always want to be able to physically observe their patients, if only to combat the problem of a sick person in denial assuring them via e-mail that she's fine. Likewise, sensitive business negotiations will probably always best be done in person. Some face-to-face situations are just a matter of personal judgment. There is a hilarious sequence in the 2009 movie *Up in the Air* in which a recently hired M.B.A. convinces her employer that it is much more efficient to lay people off online than to fly someone from place to place to do it; she then breaks down sobbing in outrage when her boyfriend texts her to suggest that maybe they should see other people.

Some communication preferences are lifestyle-related. For example, someone who lives alone may be likelier to crave more direct human interaction in the workplace. Some people simply can't stand staring at a monitor and being hunched over their keyboard for long, uninterrupted periods. Walking down the hall to actually talk with another human being gives them a chance to stretch their legs. Some people lack the confidence in their ability to interpret digital tea leaves as opposed to facial expressions and body language, and therefore they greatly prefer personal interaction.

Nevertheless, many exchanges can reasonably be conducted as well online as off, even if it results in some slight disadvantage for one party or the other (or both) and is counter to a personal preference. Therefore, you should generally think carefully before deciding to request that someone meet you in person. For an increasing number of people, being at work means being available—online—to multitask and prioritize as they see fit. Asking for that face-to-face meeting may be seen as an imposition or even as outright pushy.

However, many offliners can be hell-bent on meeting you in person. Some have (necessarily) become quite skilled at it, using techniques such as deftly avoiding providing a needed piece of information unless you meet with them or calling you two minutes early for a prearranged call in the hope that you will not yet be at your desk, which gives them the opportunity to try to make you feel guilty that you stood them up. Conversely, they might stand you up for a phone call, apologize profusely, and then offer to buy you a really nice lunch to make it up to you. If you are in the business world, you have doubtless experienced this phenomenon at least once. And if you're swamped with work and realize you're not going to make the offliner happy unless you agree to actually meet him, your understandable reaction may be to try to avoid him altogether.

We often explicitly identify offliners in our investigations to separate them from people who conduct potentially dubious activities in person in an attempt to conceal their actions. The vast majority of offliners are neither sinister nor creepy, merely a bit old-fashioned or awkward. Or just plain lonely.

If you have offliner tendencies, it is important to keep in mind that many people who spend most of their workday online have become rusty in the skills needed to actually conduct a real-world meeting—or may never have developed them in the first place. These include politely but firmly closing the meeting when it has ceased to add value, something many people find much more difficult to do in the real world than in the digital one.

The Loud Talker

This is the person who buzzes people on IM, uses lots of bold text, capitalizes whole sentences, and seems to think you just can't have enough exclamation points or emoticons. The enthusiasm of such people can be contagious, but not quite in the way they imagine.

While people with strong Digital YOUs may quietly pity the offliner, the loud talker is an archetype who is often just as shunned. He has no direct real-world counterpart. The closest thing would be someone who continually talks fast and loud at those around him, oblivious to their lack or interest and to the fact that he is showering them with a stream of spittle in his excitement. Often a power user of all things online and well aware of the infinite sea of e-mails, IMs, SMSs, and so on, he knows how to make his messages pop. The loud talker may just be exuberant and enthusiastic by nature, but more likely he is driven by a fundamental insecurity that others will not pay proper attention to him or follow instructions. Often saddled with an utterly thankless job function that many people really do ignore, the loud talker finds a way to yell using every possible form of communication. Take the following examples in different kinds of media:

E-mail
- Using different-colored fonts, often in the same sentence
- Using boldface, underlining, or similar effects
- Inserting animated images to draw the eye
- Excessively using capitalization and exclamation points or question marks
- Repeating the same text, in case the reader missed it the first half dozen times

- Resending the same e-mail, perhaps with the aid of automation, with increasing frequency as a particular event approaches
- Indiscriminately using "URGENT," "IMPORTANT," or "Please read!!!" in subject headers
- Routinely flagging e-mail as high priority

IMs

- Buzzing you repeatedly
- Doing lots of sends (Instead of composing a complete message before hitting the Send button, he hits it after every few words, creating a stream of "ping" noises to remind you that he is still communicating.)

Any home pages

- Using all of the preceding techniques, plus obnoxious widgets that autoload with the page and contain attention-grabbing—and loud—audio or video.

Unfortunately for the loud talker, the combination of these techniques accomplishes, at best, only half the mission. Having unavoidably seen the bold, blinking, bright purple text in all caps, recipients may passive-aggressively ignore the call to action that was so important to the author. Worse still, if they find the composition of messages too obnoxious, they can defend themselves in future by simply not opening the next message from the loud talker or, if he is using e-mail, by autofiltering him into oblivion for good.

The cold, hard reality is that all of the nifty text, sound, and image effects often backfire when they are overused or used improperly. Loud communication is not the same as effective communication. It never will be.

Some loud talking is purely situation-driven, the literal equivalent of someone yelling digitally. This is why our software looks for certain combinations of words and loud talking indicators such as all caps, extensive use of exclamation points, bold text, and so on. The following is a typical example:

GUYS, PLEASE STOP F*CKING AROUND WITH THIS AND ADHERE TO THE WISHES OF THE LEGAL DEPART-MENT AND LET'S GET THIS DONE!!!!!!!!!

Needless to say, this is the kind of conversation that should always be conducted in the real world rather than the digital one.

Sometimes it can be hard to avoid the habit of loud talking—especially if you are tasked with communicating about a topic no one seems to care about. If you happen to be in this particular boat, here are a few things to keep in mind:

- There is absolutely a time and place to bold, italicize, underline, or color text. But if you are doing more than one of these things in the same sentence, you should ask yourself whether you're descending into loud talker territory.
- If you feel compelled to use a multitude of effects within a particular message or document, create an ASCII (no visual effects) copy and read the text again to see whether it is clear and concise.
- While there are occasions when sending the same e-mail to the same audience multiple times makes sense, most of these involve do-or-die deadlines. So unless you have one of those deadlines or quite a long

time has passed since the last reminder, you should really try to avoid doing this. If you cannot, add new text to the top of the message explaining why you are sending it yet again, as well as any new information that might be relevant. If you can reduce the number of people receiving the second message, perhaps because some have already complied with the request it contains, do so. People always appreciate it when their good behavior is noticed and rewarded, even in small ways.

> **Reminder:** You can use the Digital Mirror software to see who brings out the loud talker in you and who is a loud talker when communicating with you and/or those around you.

The Public Shamer

We all know people who, even if we don't completely ignore them, register low on our priority list. Public shame, far more quickly and easily achieved in the digital world than the real one, is their frequent revenge. Think of the colleague who sends you an e-mail asserting that you did not respond to one of her requests promptly enough and cc's everyone you've ridden an elevator with in the last six months. Or the relative who cc's you on a message thanking another relative for remembering an important occasion that you forgot. One of the hallmarks of this archetype is that their public complaints or chastisements tend to outnumber both their private messages and their attempts to preempt a possible problem by, for example, reminding a distracted relative of an important family occasion in advance.

Group e-mail aliases, chat groups, wikis, and various community forums make it easy to publicly blast, or at least chide, someone the shamer feels has unfairly or rudely ignored her. All it takes is a few keystrokes. And it will surely conjure up appropriate sympathy from others and shame the jerk in question into behaving better the next time, right?

The problem in the digital world, just like in the real one, is that no one likes a whiner. And once someone acquires a reputation as a public shamer, the technique quickly loses its efficacy. Worse still, people may proactively shun the public shamer so as to avoid the public ridicule that will ensue when one of them inevitably fails to meet expectations.

Public shamers also can pose problems for their employers, because their frequent complaints make it more difficult to discern real problems from perceived minor slights.

The Digital Exhibitionist

We all know people like this. They are the digital counterparts of the guy in the cubicle who has managed to fill every square inch of his personal workspace with a bewildering variety of photographs, award plaques, diplomas, and other references to himself. The Internet now provides this personality archetype with all of the virtual space he craves for his memorabilia. All that remains is to coax people to come and admire this beautifully conceived and maintained projection of his digital self. The gentle nudge of an appropriate link in an e-mail signature, under a blog posting, and so on may not be sufficient to generate the volume of traffic the digital exhibitionist feels he deserves. So bolder measures, such as the following, are required:

> You've got to take a look at these great pictures from my trip: www.digitalexhibitionistsrus.com/jr32.htm. Tell me which one you think is the best because I'm going to pick one to frame as a wedding present for Joe.

This message is a double whammy: not only is the link being shoved directly under your nose, but a specific response is being requested. Furthermore, a truly dedicated digital exhibitionist will eagerly scan the log of visitors to his website, looking for a domain name that likely indicates you. If this is a one-time request, that's one thing. Certainly there is nothing wrong with asking a friend for an occasional opinion. But if it is part of a consistent pattern, you will likely interpret it as a grab for attention.

The number of people who document a significant part of their lives in detail on the Internet over a period of years is astonishing. It is, in a sense, the modern, high-tech equivalent of maintaining a diary or scrapbook. Like a scrapbook, it is usually intended for use by you and your family. The fact that someday many of these publicly accessible pages will be attacked and mined by sophisticated bots (software that runs automatically over the Internet), which will be seeking ever-more effective ways to market to you, does not yet seem to be stopping most people.

Often, the motivation seems to be linked to a belief that the more personal information the Digital YOU is willing to put out, the more people will be interested in you, at long last seeing you as that deep, sensitive, quirky individual you have always known yourself to be. That exposing every aspect of your life will make you more attractive to your coworkers, prospective employers, potential dates—and naturally that soul mate who is doubtless out there somewhere looking for you. Or that, against all statistical odds, your content will somehow gain that 15 minutes or more of fame and droves of visitors will come to pay homage at the temple dedicated to your Digital YOU.

Unfortunately, exposing everything about yourself is not necessarily a good idea in the digital world or the real one. The commonly used phrase "too much information" exists for a reason. Too much of a good thing is almost certain to be boring and has a good chance of being embarrassing as well. While Jerry Seinfeld may have the unusual gift of making the mundane details of everyday life genuinely humorous, most people would fail to generate even a single chuckle if they were to embark on a description of the Chinese restaurant that made them wait an hour for dinner. Further, content designed for one purpose, or thoughtlessly posted for laughs, can backfire when examined in a different context.

One good example of this involved a young bank intern who sent e-mail to his supervisor informing him that he'd have to miss a couple of days of work because of a family issue. The supervisor thought little of it until a photo surfaced on Facebook of the intern at a party the same night as the alleged family emergency. In the photo, the intern was dressed in a fairy costume, complete with wand in one hand and a can of beer in the other.

But the supervisor had a sense of humor. He responded to the original message explaining the family emergency with the following:

> Thanks for letting us know. Hope everything's okay in
> NY. (cool wand.)

The message, which was bcc'd to the entire office, contained the picture in question, thus outing the intern, and generating much debate on various Internet forums. (You can see the picture and e-mail trail at http://valleywag.com/tech/ your-privacy-is-an-illusion/bank-intern-busted-by-face book-321802.php.)

To take another example, long and ongoing blog entries documenting painful breakups are extremely common on the Internet. Whole websites are devoted just to breakups, such as the now apparently defunct mybreakupblog.com, where the most popular breakup blog had over 59,000 visits. Such sites provide digital exhibitionists with a means to vent their frustrations, and digital misery loves company even more than its real-world equivalent. Indeed, in this arena, Digital YOUs reach out and publicly embrace one another for their support through those first few difficult weeks. And it shows the wrongdoer just how much she is undeservedly making the digital exhibitionist suffer. Likewise, it shows others who are going through similar events that they are not alone. As pitched on the previously mentioned website,

> Q. Will [blogging] help me with recovery from my
> breakup?
> A. Of course it will! Nothing helps you understand
> and heal your inner being as much as writing down
> your own feelings and thoughts. Blogging/journaling
> enables you to release those negative feelings and
> consuming thoughts, take a step back from them, and
> then it gives you the ability to see things much more
> clearly.

This is all well and good until a particularly thorough HR person or recruiter decides to spend an afternoon rooting through every bit of available content about or by you on the Internet—or even *probably* about or by you. Perhaps by this point, the episode in question has faded in the rearview mirror. Maybe the two of you are even back together. But the wrenching and possibly pathetic description of your state of mind in those horrible weeks may not be consistent with the confident leader the recruiter is looking for to fill that key position. Consider the following example, excerpted from a real breakup blog:

> But I miss him. I want to call him but I won't. My friends are tired of hearing about him. I'm tired of thinking about him and I just wanna run away from him or WITH him. Escape this love madness, craziness, obsessiveness. It's obvious he's on my mind 24/7. I can't use the bathroom without thinking of him. I miss him. I love him. I'm strong though. I'm in the stage where I be a BIG GIRL. Don't cry! Accept the experience for what it was and let it go. Most of all I'm working on not thinking about him. I just feel so exhausted cause I've cried for the last two nights. Just about everything.
>
> LOVE—A terrible thing to waste.

Understanding the possible consequences of putting this kind of information into a public forum may just seem like common sense, and indeed it is. But one of the risks that the digital exhibitionist runs is that he has exposed so much content and so many potentially intimate portions of his life over a long enough period of time that he loses perspective on how certain parts of the content could be perceived by a new acquaintance or a stranger—and the risks that go along with this. After all, the postbreakup period lasted only a few weeks, and 11 glorious years of his existence are meticulously documented and easily available from one or more of his home pages.

Unfortunately, from an outsider's point of view, phrases like "I can't bear the thought of going on without her," and "I was too depressed for weeks to get out of bed, other than to consume more alcohol," jump off the page, leaving all else permanently obscured in shadow. Employers are generally risk-averse, and anything that might be off-putting to an employer is probably a bad idea in other contexts as well.

Note: Although many social networking sites do have different levels of access to personal content, the safe assumption is always that if the content is there and is clearly associated with you, any enterprising soul who wants to see it in full will find a way to do so. Further, it can be awkward to not provide Friends list access to colleagues who request it. A better strategy, therefore, is to restrict any content you might not want associated with the real you to specific forums where the Digital YOU can express itself entirely unfettered through the use of an alias or a user name that in no way clearly tracks back to you. However, even when taking such precautions, the more details you provide (a series of somewhat unusual first names, specific locations, and so forth), the more likely it is that you could still be unmasked.

There are, of course, many good uses for MySpace, Facebook, and similar sites. Among other things, they make it much easier for friends who are separated by geography to keep up with each other's lives. Questions of judgment are raised not by the mere use of these vehicles, but by what level of detail and type of information you provide and by how vigorously you try to drive friends and coworkers to visit—and keep visiting.

The Digital Credit Grabber

This archetype is a relative of the buck passer but merits an examination in its own right because it occurs so often. Whereas the buck passer lives in fear of taking responsibility for a decision, even a fundamentally correct one, the credit grabber is the digital equivalent of the real-world know-it-all—only far more insidious. After all, the Internet has the answer for everything, right? Surely the only thing that stops you from knowing whatever you want to know in mere minutes is laziness.

This is certainly true in simple situations—for example, just about anyone can easily find the nearest Starbucks online. In most other situations, however, two major variables confound this assertion: the varying quality of information on the Internet; and what the user already knows, which limits her ability to interpret the information correctly.

If you are thinking that the issue of quality information isn't a problem because of all the social networking sites that provide community ranking of content, think again. All these models have a fundamental limitation: accurately judging the limits of someone's expertise is, in fact, a difficult and unsolved problem. For example, let's say you are an expert in French cuisine. You can—and have—written many highly knowledgeable, accurate, and useful reviews of restaurants featuring cuisine from different regions of France at different price levels.

But what do you know about Mexican food? Your knowledge about Mexican food is not inferable from what you know about French food, even though you may write a review of a Mexican eatery with the same confidence level used in your reviews of French restaurants (because the one thing you *do* know is what you like and don't like). However, the credit grabber who takes someone out on a long-awaited first date to that Mexican restaurant he thinks is great based on the review you wrote on your favorite foodie site may be in for an unpleasant surprise.

Of course, had the poor slob now going home alone originally fessed up to the fact that he'd read about this restaurant on an amateur food critic's blog, the consequences would have been less severe. He and his date could tell jokes about the wackos on the Internet. But members of this archetype don't typically think in these terms. Rather, they want to project to the world—and especially specific people—that they have the refinement, taste, cultural knowledge, and so on to know the right place to go; they are not shut-ins who sit home on weekends surfing the Net instead of going out to cool new places.

However much social mischief the credit grabber can make, he is even more terrifying in a professional context. Although these folks are sometimes easy to detect at first blush, this is not always true. If you work in a large company, think about the last time you were in a meeting where you did not know many of the presenters. Someone gives a confident, slick, and well-thought-out presentation. You are suitably impressed. A few weeks later, you reach out to this person and ask for some information on a related topic for an important proposal that is slightly outside your area of expertise. The colleague initially exceeds your expectations by quickly replying with the requested information. All is well until your proposal is rejected because it contains some completely irrelevant garbage. Only then do you connect the dots and realize that the masterful presentation was likely "borrowed" from somewhere and, while doubtless still a good presentation, was created for some other situation, which probably bore little relation to the one at hand. One particularly disastrous variant of this scenario is when you unwittingly make use of a sales presentation, courtesy of your local credit grabber, that originated with a competitor and is recognized by the prospect.

This kind of credit grabbing is only a few stops short of outright lying, if only by omission. So when caught, the perpetrator may act a bit like a cornered animal, even if the offense is a minor one. Particularly dedicated credit grabbers may deliberately change enough words in what they copy to make it more difficult for their "borrowed" content to be detected by others.

Most credit grabbers are not evil. In a social context, they are only trying to appropriate someone else's experiences, thoughts, or clever quips in order to appear more charming or interesting than they actually are. In a professional context, most truly want to please and be seen as helpful and competent. But their eagerness to be right often eclipses their better judgment. Intoxicated by the idea that the Internet means never having to say, "I don't know," they lose sight of the fact that copying content with nuances they may not fully understand, and with which there may be copyright issues, may mean not just the humiliation of being busted, but also the possibility of getting fired.

The best way to avoid problems—whether real or perceived—is to explicitly provide a source whenever you quote or copy content or reuse materials. It doesn't matter whether the source is a wiki, a public Internet forum of some kind, or an e-mail from a colleague. By giving a source, you are not only assigning credit properly but, in many cases, alerting the recipient to the fact that the content in question was originally created for a different purpose and that the expert source is more handshakes or links away than you, should the recipient need to contact her.

The Bull in the Digital China Shop

This is a naturally careless person with an even more careless digital instantiation. The reality is that the Digital YOU is entrusted with insane amounts of information on a daily basis—not just factual information, but thoughts, opinions, fears, desires, anxieties, and rumors. This is in addition to ancillary knowledge, such as the fact that someone stayed up all night (evident from time stamps of posts, messages, log-ins, and so on, or from a reliable indicator of active online status). While some documents contain labels with instructions such as "Do not forward," "Do not send outside the company without written authorization," or "Not to be released prior to June 23," the vast majority of messages, documents, and other electronic items most people receive bears no such instruction. On the contrary, there is the implicit presumption of a universally shared understanding of what is and is not appropriate to do with the content.

However, few things in life are truly universal, and this is no exception. If a friend or colleague has ever been angry with you because you shared information she gave you, or shared it in some way that she considered inappropriate, you've experienced firsthand how much angst and anguish even an isolated instance of this behavior can generate. Those who feel they have been burned by this archetype often wear the stamp of it on their virtual foreheads for months afterward, as witnessed by e-mails that start with admonitions such as "I AM SENDING THIS MESSAGE ONLY TO YOU—DO NOT FORWARD."

Unfortunately, common sense is an even rarer virtue for Digital YOUs than it is for real people. And informal digital communication requires not just common sense, but considerable diligence to handle it correctly. Consider that, in most office environments, some people have social ties to one another as well as business relationships. This increases the likelihood of what we call *commingled communication*, meaning a single e-mail, IM, or any other kind of data object that contains both work-related content and personal content. Indeed, in our experience, such commingling is extremely commonplace. Often it is harmless and offers no opportunity for embarrassment—for instance, the message discussing a procurement request with an attached zip archive containing photos from the sender's recent trip to Hawaii.

Now let's consider an identical message about procurement, but in the postscript the author goes on to ponder what he should do about the fact that his wife is having an affair, or that he may seek a transfer because he is fairly sure that his current boss is a cokehead. Alternatively, a speaker note on a slide in a presentation might say that the slide is necessarily very basic because "the customer is, unfortunately, a moron." Such things really do happen, making the china shop analogy a good one; knowing that such breakable china (sensitive material) is all around us should make us all operate with a higher degree of caution. It is a good idea, for example, to read all the way to the bottom of an e-mail from a close friend or colleague before forwarding it to someone else. Verify that the version of a potentially sensitive document you are posting is, in fact, the correct one (and not the one with all of the user comments that say things like "Wait—this can't possibly be legal"). Be careful about adding a new participant to a two-way chat that contains some sexually explicit content. And so on. People should naturally take these precautions, but how often do they?

Even without comminglers wreaking havoc, properly tagging information—what a document is about—then segmenting it accordingly—who this content may be shared with and who should not see it—is a difficult task. Remembering all of these decisions correctly months later is far harder still, as is determining whether these decisions still are appropriate months later.

Even the most seemingly benign topics can contain content that is potentially explosive. Long e-mail threads, in particular, are often excellent hiding places for things that would look ugly if they were ever exposed to the broader light of day. This is because of the phenomenon known as *topic drift*, in which conversations often stray significantly from their original topic as they progress. We saw a great example of this in a criminal investigation involving kickbacks of various kinds. It was a long e-mail thread, most of which discussed the relative merits of different basketball teams. Unless you had a real passion for basketball statistics (or happened to be a computer program looking for specific things), it is highly unlikely that you'd have the patience to read the full thread all the way back to the start. However, the thread had originally been initiated as a discussion about whether it was advisable to give expensive season tickets to some executives from a partner company so as to help send more business their way. As the thread expanded and people started to offer to take bets on this team or that, who knows how widely it could have traveled?

If you are wondering how we caught it, it was essentially done using what is known as a "frame." We knew that certain types of things of value were alleged to have been offered as bribes, including luxury vacations, golf club memberships, and excellent season tickets for major sports teams. We had also used our software to compile a list of the names of people working for different business partners of the company we were examining. So we programmed an action to look for "give tickets" to the right kind of person, and the e-mail thread was a search hit, despite the misleading surrounding context.

More mundane examples include service or sales personnel logging complaints about how dense a customer is, only to find their comment going right back to the customer as part of the text in a problem resolution report. While this sounds incredibly stupid—and it is—the reality is that the workday is full of people asking most of us for this kind of information or that. We naturally want to oblige—and also be home in time to watch the game when everyone else does. So the urge to just pass along that mess of data without really looking at it or assessing potential china breakage is understandable. But even in purely social contexts, the opportunity for trouble exists. For example, discussing someone's involvement in some types of special-interest online forums can cause a number of problems, whether it involves that person's politics, personal fetishes, or anything to which someone else might viscerally object.

If you've been accused of these tendencies, chances are it's because you tend to see all of the e-mails, IMs, documents, and so on that you receive as one giant soup of information, in which most bits of data float around freely in the bowl. Those that aren't just floating in the soup are likely explicitly labeled with the appropriate restrictions. But the digital world is actually segmented into many different fragments that no one expects to intersect with one another, but that sometimes do. For example, you wouldn't expect to see your grandmother perusing content on a new community website devoted to the local clubbing scene; you assume she's hanging out at the online senior citizens' forum. But she may find her way to the clubbing website if there is a picture tagged with your name. Or she might be sleuthing around on an online dating site for twentysomethings and find that Rob1212, who is breathtakingly and memorably specific about the physical attributes he is looking for in a potential date, is you.

A sin of omission that the bull in the digital china shop often commits is not paying enough attention to who his digital audience is. For example, most group e-mail aliases tend to expand over time rather than shrink. At any rate, few such aliases that survive long have an entirely static membership. Many companies offer tools that can list the members of any group e-mail alias of which you are a member. This is an excellent resource to use from time to time or whenever you're in any doubt about who exactly you are addressing. Nevertheless, it is not always obvious who might have access to a particular part of a wiki or other online content.

There is a nearly infinite number of ways to get into trouble that relate to group e-mail. A common one that often leads to complaints to HR or worse involves porn or dirty jokes being sent to a group e-mail alias for a management team whose obvious members are all male. This may seem safe enough on the surface, even if it flouts corporate policy; after all, it's usually no worse than the jokes routinely told around the table during the weekly management meetings—except for one small detail. In many companies, the managers' female administrative assistants, who do not attend those meetings, *do* receive those e-mails and may not be amused. The potentially fatal scenario is the one in which such an invisible member of the alias is actually the target of inappropriate words—or worse still, pictures.

To combat bull-in-a-china-shop tendencies, you must first consciously define different categories or segments of content you receive. The breakdown can be hierarchical, if that helps: for example, a friends category could be further broken down into basketball friends, work friends, college buddies, former boyfriends, and so on. You might also do this by topic or category of item, such as a party invitation or a church activity announcement. Once this is done, whenever you are about to transfer content from one category to another, take a moment to consciously ask yourself whether harm could come from it. When corresponding with someone who is a member of more than one category, be particularly vigilant as to which category you are communicating about so as to avoid inadvertently crossing the boundary— revealing confidential work information to your basketball buddies, for example. Repeat this exercise for specific situations, such as with two of your friends who can't stand one another, have just broken up, or have wildly opposite political views. These kinds of precautions will help prevent awkward situations—such as the dumpee seeing that you and the dumper are heading to a party a mere two days after the breakup.

The Last-Word Getter

Members of this archetype don't allow anyone other than themselves to have the last word. Essentially, they have to be right the whole time, every time. Always eager to make insignificant—and public—corrections to someone else's e-mail, document, or posting, or to extend a discussion when it's clear that few others are still interested, they are insecure people who are constantly trying to demonstrate their expertise or value.

The digital realm adds an interesting twist to this real-world personality type: unlike in the real world, it is easy to adopt more than one online identity, all of which can project the same viewpoint. This makes it much easier to get in *more* words as well as the *last* word and also to establish greater credibility by creating a false sense of others who share the same opinion.

Public Internet forums offer interesting examples of this kind of behavior. On groups like the Yahoo! Finance message boards for different companies, the same Digital YOUs are often duking it out to win the argument or join the prevailing viewpoint. Often these are not real Digital YOUs and cannot easily be traced back to a real person. User names and e-mail addresses on community sites are freely available, so genuine cravers of the last word need not be limited by having only one digital means of projecting their views on the world. Even IP addresses can be spoofed or concealed in an attempt to shield their location and/or identity. In short, hiding is very, very easy.

Why hide? Sometimes the real person has inside information about the company in question and therefore wants to remain anonymous. Or perhaps he is a former executive of that company and doesn't want anyone to perceive the axe he is grinding. Or he may fear a lawsuit for libel. Or all of the above.

By using a technique known as linguistic fingerprint analysis, we can often determine that several Digital YOUs probably correspond to the same last-word getter. Linguistic fingerprinting refers to the practice of identifying the author of content based on specific writing quirks and habits. This includes everything from certain types of spelling and grammatical irregularities to a fondness for certain words and expressions, punctuation, sentence structure, and so on. While such methods are not conclusive, when restricted to any given population of people, they are often highly suggestive. And they offer the advantage of assessing things that people are unlikely to consider changing and that would be hard to change anyway, because such habits are established for most of us in grade school.

Sometimes the desperation for the last word is not a global personality trait but simply a strong reaction to a particular individual who really gets your goat. Some Digital YOUs, for example, rarely if ever initiate conversations as opposed to responding to the comments of others—often very specific others. It is therefore not uncommon to see digital identities who only manifest themselves in order to negate comments made by a single individual. Such situations appear to involve one individual attempting to pummel someone he dislikes via a coordinated attack with multiple identities.

Last-word getters rarely do anything more serious than slightly annoy people and, on occasion, perhaps make themselves look a bit ridiculous. However, their behavior may discourage others from entering into dialogues with them if there is a way to avoid it. Remember, as in the real world, not every digital argument is really worth winning, and not every minor point is worth making.

It should be stressed that in the digital world, just as in the real one, there is a vast difference between occasionally doing the ill-mannered, inconsiderate, or ill-considered thing and doing these things as a matter of course. No one is perfect all the time, whether online or in the flesh. What you do most of the time when you interact with someone is what really matters.

5

The Digital YOU in a Bad Mood

IF YOU'VE BEEN in the same work environment for several years, you've probably noticed that everyone hits a rough patch at one time or another. And even without life-altering personal drama, we all have the odd bad day for mundane reasons ranging from getting a speeding ticket to suffering from a sinus headache.

In this one arena, the Digital YOU is at a disadvantage. Imagine that you come into the office with a hacking cough and runny nose. Most people will physically avoid you and will likely be understanding if you are slow to respond to a request or miss a minor deadline. By contrast, if you are working alone from a remote location, even if you are coughing up science experiments, the need to react is gone, and the overall impression is not as vivid. You aren't going to get as much extra slack—maybe none at all.

It is easy in person to recognize when someone we know well is depressed: the slouched shoulders, the lethargic movements, the glassy stare, and the clothes that look like they've seen the inside of a laundry hamper but not a washing machine. Even the fairly unobservant among us can detect subtle changes in body language and other signals that suggest all is not well in someone's world.

In contrast, the Digital YOU is largely unencumbered by physical or other real-world problems, because so many digital relationships remain constrained to a single context, whether it is a shared personal hobby or a work-related issue. In an actual office environment, you are expected to ask your cube neighbor about his children occasionally out of politeness, especially if his walls are full of their pictures. If he is constantly on the phone with family members discussing his mother's illness, you can't help but overhear. If he is about to get married, it is impossible for you not to know about it. So even if you don't really care, to a large extent you have no choice but to opt in to some degree. Not all aspects of your neighbor's life will be revealed, but over time, one way or another, a certain number of them will be. And from this information, you will easily come to recognize those days when it is best just to leave him alone.

This isn't to say that you probably couldn't easily amass much the same data about him via Facebook or the like. But there's a key difference: you have to actively take it upon yourself to keep up with what's new on a regular basis. It's not like those pictures of children staring out at you whenever you go to recaffeinate. Further, the information presented online is likely to be positively skewed relative to what you could directly observe from your desk. For example, someone might frequently argue with her boyfriend on her cell phone, to the annoyance of her officemates, but not change her relationship status online. Her home page would contain many smiling pictures of her with her boyfriend doing various fun things, but there would likely not be a single reference to any kind of argument. After all, making such squabbles visible to 832 online Friends somehow would seem like a major escalation of the fight—even if 825 of them aren't really her friends.

Most people who take the trouble to maintain their online personas try to make these personas more interesting and fun than they actually are. Almost everyone wants to appear to have a more glamorous and better life than they really do. This is why, for example, so many people post a disproportionate amount of pictures from their vacations; the vacations account for a small fraction of their time per year, but you'd never know it from their websites. Conversely, if you have an unpleasant, mundane, ongoing aspect of your life, such as a chronically ill family member, you probably can't hide it forever from those you spend 40+ hours a week with in the real world, but you are unlikely to broadcast it far and wide. (For obvious reasons, many users in emotional support communities use aliases such as "tornandbleeding" that bear no relation to their real names.)

Changing his public relationship status aside, cues that someone's digital persona is having the equivalent of a bad hair day are often subtle when compared to his real-world counterpart. In most cases, they are restricted to the following:

1. **Attempts to avoid or at least delay interacting with people who irritate you.** This takes on many forms:
   - Communication that contains delaying language, such as "I'll get back to you on that," "I'm really underwater today," and so on. The person in question may actually be amazingly busy, but even so, she is more busy for some people than others. *Note*: the Digital Mirror software can help you see with whom

you typically use such delaying language and who uses it with you (in the Blow Off Scoreboard).

- Setting statuses indicating that a person is away from his desk despite the fact that he is physically there.
- Buck-passing, even if it is not generally the person's habit.
- Simply ignoring e-mails, IMs, and other communication attempts from certain people—but not all people.

In the real world, it can be very difficult to avoid an irritating colleague if the clod happens to have a cubicle a few feet from yours. In the digital world, there is the seductive illusion of being able to hide from that obnoxious coworker. But it is just that: an illusion. Consider the coworker who IMs you to let you know that he doesn't have the time to talk to you about something today because he has to leave in half an hour to pick up his son from soccer practice; you then discover a continuous stream of e-mails, document check-ins, updates to information in a Web application, and so on from him during this time.

2. **Less attentive or less detailed responses than normal.** A good example of this is when someone who usually replies in-line or point by point in e-mails just writes his response at the top. Not only does it take less effort, but it can mask the fact that he didn't respond to all of the points. The same thing goes for someone who normally goes in and edits a document as opposed to commenting briefly on it in the cover e-mail. Especially when these things are done rapidly, it is likely not a sign of the person being truly busy as much as merely wanting the task to go away—and the sender along with it. (Consider that if you are monumentally

busy with some critical task, you are unlikely to stop to read a document or respond to a longish e-mail.)

Another example is sending someone a link to content rather than a written response. Hopefully, the desired information can be found in the destination page, and if not, it will likely confuse the recipient enough to slow him down until your throbbing headache disappears.

There is also the time-honored strategy of delaying or diffusing a question by responding with another question. Some people do this as a general matter of communication style to help train coworkers to ask specific, well-formed questions so that a lot of back-and-forth is avoided. But if someone does not normally do this and sends a cluster of such responses at one time, it is a short-term avoidance-of-nuisance mechanism.

3. **Slower-than-usual turnaround times, including not showing up when expected.** The thing about observing deadlines or punctuality for meetings—regardless of whether they are in person, via telephone, or online—is that those people who care about such things are rarely late unless something is really wrong. Normal patterns can be determined from electronic records in a number of ways, including analyzing meeting minutes and schedules, as well as the subject lines of e-mails such as "Why weren't you at the meeting?" or "Where are you?"

4. **A change in overall tone to being noticeably more abrupt, possibly crossing the line into rudeness; alternately, being unusually sweet as a countermeasure against abrupt behavior.** Personal communication style is something that, under normal circumstances, tends to change very slowly, if at all. We each have our own style, as well as

our own measures of what is appropriate in different contexts. Most people have a formal mode that they use with strangers and/or those whose status is in some way superior to theirs. But otherwise, given the sheer volume of messages that most of us generate in a day, most communication is ad hoc. Put it this way: if you only had to compose a few messages a day, you'd likely give each message far more thought than you do when you have to compose hundreds. As it is, few of us have the luxury of drafting most of the messages we send each day, reading them over once or twice, and making a few well-considered edits. We just couldn't keep up in the modern workplace. Instead, most of us write electronic messages with a level of spontaneity that is more similar to speech than to writing a college English paper; just as we rarely rehearse what we are going to say before we say it, we type and hit Send. Both the initial thought and the articulation of it happen in a flash.

When tone changes, it is usually a sign that someone is too tired or stressed out to maintain the minimal veneer that most people reflexively exhibit in a professional context. The change can be obvious, in the form of rudeness, ridicule, or sarcasm. Further, if someone fears that she is on the verge of biting other people's heads off, she may overcompensate with exaggerated politeness. For example, someone who doesn't usually load down messages with many instances of the word *please* but uses it eight times in one message is almost certainly irritated.

5. **A seemingly inappropriate focus on a particular topic.** Is someone doing lots of tweeting about various vacation ideas the day before a deadline that, if missed, will all but ensure that he'll have plenty of vacation time available? Digital YOU daydreaming is a natural and even potentially

healthy way to diffuse stress, but like regular daydreaming, if you spend too much time doing it, it becomes a problem rather than a healthy release mechanism. Further, if you get caught waxing eloquent about the merits of particularly spectacular beaches, it is much harder to claim that you were actually thinking really deep, important thoughts than if you had just been staring out the window. Saying, "I just drifted off into tweeting" doesn't sound so good.

Even if the cues of digital irritation are subtle, the extent of a bad mood is absolutely no different than it is in the physical world. There is no getting around the fact that if the real you is in need of repair—or just completely out of commission—it will affect the Digital YOU by exacerbating natural tendencies and resulting in social niceties falling by the wayside. The following true story may be extreme, but it vividly illustrates the point.

The Death of a Two-Headed Norse Hacker God

If you'd ever met the amazingly brilliant Norwegian hacker Erik Naggum, either digitally or in person, you'd never forget him. In person, he was charming, witty, and handsome, the polar opposite of the image that is generally conjured up by the term *hacker*. However, by dint of the fact that he spent most of his life in Norway, relatively few people with whom he interacted in cyberspace ever had the pleasure of actually meeting him face-to-face.

Erik is something of a legend in certain parts of the computer science community. His online persona can be best understood as a digital Batman: unrepentant and vengeful in his pursuit of criminals. Except in this case, his prey were

sloppy, unenlightened, or hopelessly mediocre programmers conversing on technically oriented online newsgroups.
Erik did, indeed, have near-superhero powers. Specifically,
he could insult someone online in such a perfect way that
the target risked becoming forever known by Erik's eloquent
barb rather than by his real name.

Some of his targets actually did mend their errant programming ways, and some will assert that Erik—or more
precisely, fear of Erik—made them better programmers,
better spellers, or even just better people. Others, such as
the poster of the following comment, managed to maintain
a balanced view:

> I remember I was flamed by him one of the first times I
> used the News system in TOPS-20 (DEC) back in 1986 at
> the University of Oslo. It took me some time to recover.
> Behind the rough usenet persona, I remember him as
> a nice and a very smart guy. We took one program-
> ming class together (IN110) at UiO, and he became a
> legend among his fellow students, as he solved the
> dreaded "The Eight Queens Problem" in 10 minutes,
> while most of us struggled to wrap our minds around
> this recursion thing for days to solve the problem.[1]

Unsurprisingly, many others were resentful and angry—
permanently. The Internet allows everyone free speech, but it
can never make everyone equally eloquent or worth tuning
in to read the rant of the day. Many people tuned in to Erik.
Not only could he write software programs faster and better
than the next guy, which brought him credibility and admiration, he could insult others back to their ancestors with
just a few keystrokes from as far away as Oslo. In multiple

languages. Ouch. His often-over-the-top remarks became a guilty pleasure for many, a sure source of a chuckle, or relief that he wasn't coming after them.

The shocking, almost schizophrenic difference in personality between Digital Erik the Avenger and Erik the mild-mannered guy next door with the ready smile was generally ascribed to the fact that his extreme reactions were to badly written code or poorly argued logic in the abstract rather than to the essentially anonymous offenders themselves. His online persona on certain technical topics achieved such stature and such a global following that it could reasonably be argued that he really was cleaning up his part of the online universe, something from which he derived evident satisfaction. One thing is absolutely certain: Erik did not suffer fools lightly. It was a fundamental part of his nature. But his already very limited tolerance for those he considered intellectually bankrupt was reduced even further by tragic circumstance.

While Erik's case is extreme in terms of his degree of anger and intolerance, as we'll see, a far lesser, passing version happens to us all at some point. Have you ever come to work with a bad headache, only to find that an inept coworker has yet again misspelled the name of an important client on a brochure, that the wrong material has inexplicably been ordered yet again, or that the worst temp you have ever encountered is once more enjoying a post-doughnuts snooze in the vacationing secretary's chair? Did you find yourself wondering how much trouble it would be to have the perpetrator of such repeated stupidity fired? But then you likely took a couple of aspirin and reminded yourself that you probably couldn't get rid of all the offenders. And with the realization that you have to see these people five days a week in the real world, at least for the foreseeable future, you decided just to forget about it.

Now imagine you know that you are frequently going to be in considerable pain for the rest of your life, which will now likely be much shorter than you had planned. Despite your daily herculean efforts, you see the same repetitive stupidity day in and day out. But you can't fire the bums or even give them the face-to-face telling-off they so clearly deserve. The fact that it is at least a somewhat different set of bums each day changes little; the effect is still that of a digital version of Chinese water torture. There's no need to conceal your frustration in order to keep the façade of peace, because it's not your job or your paycheck, but just a hobby. Plus, you are great enough at what you do that you'll never be in want of a paycheck, no matter what you say or how you say it. So there is no motivation to grin and bear the stupidity as opposed to trying to do what you can to improve matters. Now perhaps you can begin to imagine how Erik the Avenger came to be.

When news of Erik's untimely death first hit the Internet via—of all things—a tweet from someone who noticed his online absence, it quickly became apparent that he had been suffering with a very painful disease that ultimately caused his death from internal bleeding just after he turned 44.[2] Though, in recent years, his website had indicated that he had gone into partial retirement due to unspecified health issues, the nature of his illness was not generally known. Erik might complain bitterly about poorly crafted prose or software programs, but he wasn't one to complain about health problems, even severe ones. That just wasn't part of his Digital YOU, and it was important to him to keep it that way.

Immediately following Erik's death, a few people openly expressed not only a good riddance sentiment, but in some cases went so far in tweets and other media as to say that they were glad to know he had really suffered. (Talk about a statement that no sane person would ever want associated with their Digital YOU! Imagine a potential manager seeing that you were overjoyed to know that someone who had criticized your work five years ago had died young after years of enduring gut-wrenching pain. Your résumé would promptly be put in the digital "Do not interview again *under any circumstances*" pile.) However, with the passage of several days, eulogies began to appear, and people from all over the world mourned his loss. As of this writing, a few months after his death, several Web pages have been set up in his memory to archive both his technical work and his writings on topics ranging from philosophy and politics to programming.

So in the end, the digital view of Erik came to more closely resemble the real-world one. Certainly, many people whom he mercilessly flamed must have been shocked. For years, they only saw the angry face of Erik, the wrath of a Hacker God brought down from on high to publicly smite them. None of them could have imagined at the time that his Digital YOU was a reflection of someone in a lot of pain, pain that there was no need to mask as he sat in front of his computer in the confines of his living room. Nor did they know that he was getting angry at poorly written code rather than at them personally. It is totally understandable; the possibility of such diametrically opposed personas as Erik the Avenger (Digital Erik) and Erik the guy next door (real-life Erik) developing over the course of many years is an emerging phenomenon.

> Act from reason, and failure makes you rethink
> and study harder. Act from faith, and failure
> makes you blame someone and push harder.
> —*Erik Naggum (1965–2009)*

You may be wondering why anyone would care about people being grumpy in an investigative context. For one thing, any significant change in behavior immediately draws our interest. If someone who is normally calm and polite in his communications suddenly starts using language that strongly suggests aggravation or worse, something has likely gone awry. It is our job to determine whether that something has any connection to what we care about. It's possible that he is just in an ugly mood because he has a bad head cold, so our software looks for common references to such problems. But perhaps he is seriously stressed out because of severe personal financial problems. In fraud cases, that fact is of great interest to investigators, because it increases the odds that this employee is involved in the wrongdoing. Perhaps the stress comes from the suspicion (or outright knowledge) that something untoward is going on in his department. Or the cause of his bad mood can be as innocent as an interaction with a particularly irritating individual or event.

The truth is that some people spend much of their lives in a foul mood. It is simply their natural state, even when they're perfectly healthy and well rested. Others seem to gradually sink into such a state over time as a result of being in an unsatisfying job for 40+ hours a week. Because employees in this category are more likely to see their interests as being distinct from those of their employer, they are at least somewhat likelier to do a variety of things that are bad from the employer's point of view. For example, so-called whistle-blower cases arise whenever someone—usually an employee—reports a corporation's violation of a federal law to the government. To encourage such behavior, the government handsomely rewards the whistle-blower with a significant percentage of whatever penalty it extracts from the corporation. In large cases, this can literally be millions or even billions of dollars.

By contrast, genuinely happy employees usually don't want to do anything that threatens their pleasant existence. We have yet to find one case in which a contented worker blew the whistle on an employer, and I doubt we ever will. Of course, throughout the course of an investigation, the whistle-blower's identity is generally kept secret. But trying to guess the person's identity is very important, because that knowledge helps the attorneys for the corporation understand what story the whistle-blower is telling and what types of evidence—real or otherwise—the employee might have handed over to the government.

In one such case, an initial analysis of electronic data revealed that a recently hired executive was becoming increasingly bitter as a result of clashes with the rest of the management team as to what kinds of process improvements were appropriate for the company. Communications between him and the rest of the team gradually became fewer and noticeably more strained as a result.

While that naturally caught our attention, something else was even more striking. Toward the end of his employment at the company in question, he used some of the extra time he had on his hands (given that he was no longer communicating with any of his colleagues) to start pricing yachts.

Given that he lived in California and had a base salary of less than $200,000, either he was expecting to inherit a large sum of money in a few months, or he was anticipating some other kind of large windfall. I would lay pretty good odds on it being the latter.

One of the many interesting aspects of the case was that the whistle-blower almost certainly lied to the government. He claimed that massive billing fraud was afoot; however, there was a fair amount of evidence that he knew this to be false. For example, there were e-mails in which he complained that not enough hours were being billed because of faulty internal reporting processes. Of course, there is no law against billing the government for working too *few* hours. The key point here is that the whistle-blower knew that if the government went looking, it would certainly find irregularities in the company's billing practices. If he were lucky, the government would decide that it was intentional fraud as opposed to incompetence, which is not a jailable offense, and that yacht would become a reality. But in a less-beneficial scenario, the company, fearing a prolonged battle, would settle rather than fight, and he'd still get something. The latter is what happened; he got something, but more sailboat-sized.

This one example should not be taken to mean that many whistle-blower complaints aren't totally legitimate or at least made in good faith. (Unfortunately, the letter of the law and common sense often part ways in the real world, with the result that what seems like a violation of the law actually isn't.) But the point is that once people are pissed off and feel wronged, they may look for ways to "get theirs."

Both former and current employees can be called as witnesses in any kind of legal proceeding against their employer.

They can also come forward on their own to volunteer assistance to the opposing side if they are sufficiently disgruntled. Further, sometimes a particular manager draws such an intense degree of hatred that people who have nothing against the company generally will go well out of their way to harm her specifically. Whether there is such a disgruntled employee in the mix can be an important element in determining whether it is better to try to settle the case rather than go to court and fight.

But there are also many non-litigation-related reasons for companies to care about whether their employees are unhappy. For example, in tough economic times, various types of fringe benefits are usually the first casualties. After all, no one actually *needs* bagels delivered to the office on Fridays. However, the extent of employees' negative reaction to being nickel-and-dimed in such a fashion may exceed the value of the money saved as a result, due to a loss in employee productivity. By analyzing such employee sentiments on specific topics in the aggregate (that is, without revealing the identities of the individual employees), corporations can make wiser decisions on how to spend money on their workers.

Digital grumpiness is a misery that spreads and is amplified at a frighteningly faster pace than its real-world counterpart. This also means that rooting it out takes that much longer. What's arguably worse still is that the objective size of the real-world grievance often has no apparent relation to the amount of noise that is generated in cyberspace. Knowing who the most effective transmitters of misery are is often of interest to employers. Good managers generally know who the complainers are based on direct observation. But what they sometimes don't know is how far different employees' complaints reverberate in cyberspace. Recognizing the people who have the ability to attract and influence followers is important; as managers, we'd like them working with us, not against us.

For example, in an era of mass layoffs and pay cuts, the problem of the disappearing bagels may seem insignificant. But imagine a particularly creative—and digitally savvy— bagel store owner who decides that he not only doesn't want to lose that piece of business, but he wants to deter anyone else from thinking of his wares as a luxury. So he makes a hilariously funny video of dozens of animated bagels leaving the building, being chased by an army of employees carrying sacks and holding up "Acme Corp: Bring Back Harvey's Bagels!!" signs, entirely emptying the office in the process. He then posts the video on YouTube. It hits a nerve and is viewed by some crazy number of people in a short time. Even if the bagels are quickly made to reappear at Acme Corp, the taunts will probably continue for months.

Yet another big reason for employers to care about worker satisfaction is that processes that generate significant amounts of friction generally go by the wayside. Most people don't like confrontation and will do whatever they can to avoid it. Some processes almost universally annoy people through sheer bureaucratic stupidity, resulting in a revolt of noncompliance. However, businesses usually put new processes in place to address a particular need, often to prevent the recurrence of an identified problem. So if such a process is discarded as opposed to adjusted to make it work better, whatever the bad thing was that happened before, it is likely to happen again. Therefore, identifying processes that raise employees' temperatures so procedures can be revised is a worthwhile exercise.

To me, the most interesting aspect of analyzing the various moods of the Digital YOU is understanding what types of things upset each of us and to what degree. Even the most implacable among us has at least one or two things that really get our goat. It is just a question of how often the irritants show up in our environment to cause mischief. What becomes clear, if you have a few years' worth of someone's electronic data, is the fact that some people, topics, or events reliably act as Pavlovian stimuli. To make matters worse, things tend to pile on: Your favorite team loses in the finals, *and* you've just been volunteered to spend two days with the client from hell. Still, with a large enough amount of data—and hence, a large enough number of examples of bad days from which to choose—we can see what things really set you off.

In an investigative context, we tend to look for bad moods that crescendo over time, because this is consistent with the type of stress associated with guilty knowledge. Most of us, most of the time, get annoyed one day but are fine the next. For example, you may be in a bad mood when you first discover your unpleasant assignment of babysitting the client from hell. You may be in an even worse mood during those two days. But a week later, it will likely already be water under the bridge. Many annoyances disappear from your consciousness within hours. However, any type of ongoing irritation, such as someone else getting a promotion that you feel you deserved, is likely to cause unhappiness that lasts much longer than a few days.

What types of things truly set someone off is probably as good a test of personality as any. Some people will get very upset about any breach in protocol, just on basic principle alone and even if it has absolutely zero real-world impact. Others will be made happy or miserable solely based on the real-world outcome; they won't react to the original intentions or the path taken to get to the outcome. Certain people will all but obsess about any perceived injustice or stupidity, regardless of its magnitude or the identity of the victim; some react forcefully to any perceived criticism of themselves but are largely unresponsive otherwise. Still others are unlikely to show much emotion about anything work-related, because they are apparently conserving 100 percent of their emotions for their families or a hobby of some kind.

What people often don't consider is the fact that the digital imprint of misery makes its duration and degree much clearer than it would be in the real world. And others—not just their employers—are likely to judge them on that basis. Part of the reason for this is that in the digital universe, dates are so readily associated with content. Someone can easily find out when you first started talking or complaining about problem X. It may seem like it has been going on forever, but seeing how long *actually* has lasted is often somewhat shocking, because people's perception of how long something has been an issue is often more closely linked to the number of complaints about it rather than the number of weeks, months, or years that the issue objectively existed. Further, while most people consider it a friend's duty to offer consolation after a breakup, the loss of a job, or other such traumatic events, even the most patient pal might turn pale if he saw that you had complained about the same thing 1,724 times online.

Note: You can use the Temperature Gauge view in Figure 5.1 from the Digital Mirror software to see what things set you off more (or less) and for how long, as compared to the reactions of those around you.

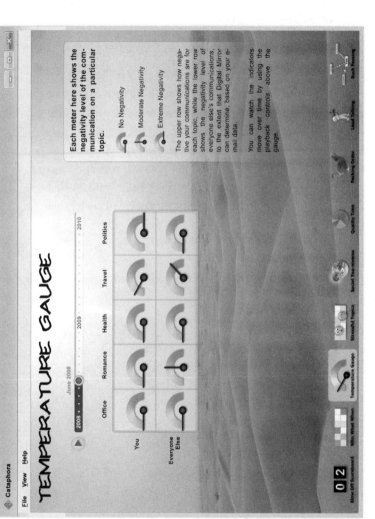

Figure 5.1 Temperature Gauge view (*Background photo courtesy of J. Gremillot—http://bit.ly/94HZEB*)

This isn't to say that the Digital YOU shouldn't express itself even when it's in a crappy mood. The key is in understanding that the permanence of the sentiments being expressed can easily make the Digital YOU seem whiny, even if that is an accusation no one would ever level at the real you. Repetition, especially, is something to avoid in this regard. Not just repetition in the same online forum or community, but everywhere. (Remember that the more information you provide on the same topic, even if you don't use your real name, the likelier it is that you will repeat details such as locations, time frames, and particular phrases that will all point back to the same source if analyzed correctly.)

6

Love, Sex, Romance, and the Digital YOU

IT IS IN the area of romance that the digital world and the real one part ways to the greatest degree. On the one hand, the digital world can actually facilitate the creation of emotional intimacy in ways the real world does not. It somehow seems safer to go further faster—after all, as they say in the movies, "we were just talking." Surely, there is little harm in that. Increasingly, the digital world seems to do everything possible to make it easier for us to talk and talk. And talk.

The amount of contact that two people who don't live together can have within a short time is now far greater than at any point in history. Those BlackBerrys, Treos, and iPhones increase the percentage of time you can reach the target of your desire through e-mails, texting, or voice. Even if the person is asleep, you can still send messages in various ways without disturbing her. If she is awake and not in a meeting, there's the ubiquitous cell phone—or free Skype-calling. In our investigations, we have seen as many as 300+ messages being exchanged within mere weeks of the initial meeting between the two people involved, just as cocktail waitress Jaimee Grubbs claims to have exchanged with megastar golfer Tiger Woods. And that's just what was still on one person's work computer many months later; there may have been twice as many messages or more that weren't saved or that were exchanged on a home machine or a cell phone.

This 24/7 blasting of messages is a highly effective way to indicate you are thinking about the other person all the time and to continuously stroke his digital ego. Simply put, it is digital foreplay. And like all foreplay, it either peters out fairly quickly or intensifies, ending up in a sexual encounter. You've almost certainly been witness to this behavior, even if you've never been a direct participant. Just recall the last time you were in a meeting and saw someone beaming radiantly while holding a partially shielded iPhone and manically texting away, completely oblivious to the goings-on around him. (Think Tiger Woods texting one of his mistresses on Thanksgiving Day, which is how his wife busted him.) You wonder whether you should leave the room or the other person should!

Sustaining that intense level of digital contact indefinitely is very hard. One telltale means of separating overt flirtation from friendly banter—such as one coworker thanking another for a job well done and saying something like, "I could just kiss you"—is its constancy over time. Most people soon run out of clever things to say at the rate of hundreds of messages a week—something to think about at the beginning of a relationship before plunging into a marathon of texting. Much of the public reaction to the Tiger Woods text messages seemed to be their emotionally immature and not terribly articulate nature. (Not everyone can equal Mark Sanford, governor of South Carolina, who more than held his own in this regard when e-mailing his lover, with such phrases as "You have a level of sophistication that is so fitting with your beauty" and "Please sleep soundly knowing that despite the best efforts of my head, my heart cries out for you."[1]) Also, there are only so many hours in the day, thus limiting the number of potential romantic prospects with whom one can exchange large volumes of text messages. One thing is sure: the pace of a relationship—including the determination that there isn't going to be one—is accelerated.

This kind of rapid acceleration can be especially dangerous in a workplace setting. Despite the obvious hazards of dating a coworker, many people do it. According to a CareerJournal.com poll, 40 percent of respondents admitted to having been romantically involved with someone at work.[2] I suspect the real-world figure is higher still. For many people, the workplace is the only real-world community they have after graduating from college. Such a community provides the inherent sense of comfort that comes from seeing how another person interacts in a common environment day in and day out. Beyond that, certain types of high-adrenaline work environments, with their many late nights, intense pressures, and high stakes, greatly facilitate such bonding—in part because coworkers usually remain in frequent communication even when they're not in the office. Such companies include start-ups, as well as any number of jobs that involve life and death for people or companies.

Before I go any further, it might be useful to explain how the romantic habits of the Digital YOU intersect with our investigative work. Our concern is to understand the character and motivations of specific individuals, so we can understand how they would likely have behaved in a given scenario. For example, a scorned ex-lover is likely to paint an unflattering—and perhaps even untrue—picture of the actions of his former partner. Lawyers and investigators can get distracted by such tainted testimony during depositions, especially if they have no clue the relationship existed.

In writing this chapter, I came to the somewhat disillusioning realization that we've never been asked to try to determine who is in love with whom, only who is sleeping with whom. The great irony in this is that someone deeply in love can certainly behave in unpredictable ways in an attempt to shield the loved one from a perceived danger, even if the object of affection has no idea of the attachment. People having a mostly physical relationship that lacks a deep emotional underpinning are less likely to put themselves on the line by risking their job or even perjury charges. They are, however, highly likely to be what I'll call "helpfully forgetful" if they fear anything bad might befall their partner.

Having the ability to evaluate true love on the basis of digital evidence would thus be highly valuable. Unfortunately, we have no technology that would allow us to do so. I very much doubt that anyone else does either—or ever could. It is inherently impossible to measure the existence or degree of love, or when it begins or ends. Love is a total abstraction; it is not a set of real-world actions, but a motivation that influences these actions. Machines can't properly detect or assess it, and people are also notoriously lousy at such assessments, even when they are direct participants. Hence there is the all-too-common difficulty of distinguishing lust from love, or of people questioning later whether they were really in love or just gave in to infatuation or the desire to be in love. Someone may not notice that someone else in her circle has fallen in love with her—or vice versa—

until jolted into the realization by some precipitating event, such as the departure of the person in question or the start of his involvement with someone else. A computer program—which of course is designed by people—thus has no chance of sorting it all out.

By contrast, identifying the existence of a sexual relationship, or at least a high probability of one, by automated examination of electronic evidence is relatively easy and always on the menu in an investigation. It is obviously an imperfect measure of romantic love, but it is the least bad one there is.

For example, here are a few phrases that, by themselves, are highly suggestive of the existence of a sexual relationship, completely independent of any surrounding context. Note that these examples are not always explicit, either sexually or romantically:

- "I left _____ at your place last night."
- "Do you want to go away this weekend?"
- "Last night was really fantastic!"
- "I'm going to wear that _____ you really like."

Or, when it is all over:

- "Leave me alone."
- "Go back to your wife!"
- "Why did you have to throw me away?"
- "I still love you."
- "I still want you back."

Similarly, detecting evidence of initial attraction or desire is usually not terribly difficult. With an ever-increasing number of ways to reach out and touch someone, it is no wonder the Digital YOU is not particularly subtle in matters of the heart. Indeed, it usually leaves behind a mile-wide trail, because digital attraction changes our normally constant habits. Suddenly there is someone with whom a treasure trove of communication—via different means, at different times of the day and night—quickly amasses, sometimes without the real you having fully registered what is transpiring. This simply means the Digital YOU can really fall for someone, or attract someone else, before the rest of you does. This can quickly lead to a dangerous situation in which the Digital YOU may overcommit the real one.

This is easy enough to do. For one thing, flirtation is actually quite complex. It depends largely on the correct interpretation of many physical cues that are totally absent in text-based communication. So the relationship may not be going as well as one of you thinks. For example, a friend of mine was rejected by a potential date online after several exchanges of messages when she clearly didn't understand that something in one of his messages was intended as a joke. Had the interaction been in person, his tone of voice, facial expression, and body language would have clearly signaled this, just as hers would have indicated immediate discomfiture, giving him a chance to right the situation.

Linguists will tell you that verbal and written communications often have somewhat different purposes and usually have different characteristics. For example, verbal communication generally uses shorter, more common words than does written communication, in which there is usually at least fractionally more time to craft your message. Verbal communication also frequently contains many types of markers that simply don't exist in written communication, such as the occasional "uh, huh" or "yeah, right" to demonstrate that you are still paying attention—sort of. And, of course, in speech, there are many cues that just don't exist in writing. What linguists call *prosody*—which refers to the duration, pitch, and intensity of speech—and intonation are used to indicate and correctly interpret the intended meaning, such as separating sarcasm from sincerity and real anger from mild annoyance.

This doesn't mean people don't try to inject more speech-oriented conventions into text, especially short messages such as IMs, SMSs, and tweets. This can be done in a variety of ways, using punctuation, emoticons, and loud talking (the latter is covered in detail in Chapter 4). Sometimes people try to inject a form of prosody by duplicating letters, for example, "oooooohhhh." However, if you've ever seen the Clint Eastwood movie *Gran Torino*, you know that writing "Get. Off. My. Lawn." doesn't quite capture the moment, the real danger that Clint might very well shoot the marauders on his lawn—and relish doing so.

But it isn't just a matter of missing cues. Let's be honest: physical attraction usually plays a significant role in romantic love, and it can be notoriously difficult to accurately predict on the basis of online interactions, photos, or even phone conversations. Often the attraction is greater on one side than the other; the ability to decide to fall in love based on compatibility, friendship, or even convenience also varies significantly. Hence the common device in movies of the kiss that doesn't feel like it's supposed to, often enabling the protagonist to finally realize that his one true love is someone other than the woman he expected it to be. Conversely, in the classic movie *To Have and Have Not*, the first movie Humphrey Bogart and Lauren Bacall ever made together, the chemistry between the two is so intense that it jumps off the screen—despite the fact that Bogie was considerably older than Bacall and not classically handsome.

Of course, you don't need to be a movie star to experience the mythical spark; you only need to be extremely fortunate. In this context, I often think of my advisor in college, Professor Roger Lyndon. In many respects, including his physical appearance, he was the typical mathematician: always a bit rumpled and disheveled, but a kind, generous, and brilliant man. He would invite me home to dinner from time to time, where his wife, Margaret, would cook us a nice meal. The two met well into their forties—or perhaps even later. He was already married at the time, not unhappily; I believe she was married to someone else as well. But it was love at first sight, and that was that—nothing else mattered, and no obstacle was allowed to stand in their way. They were still unabashedly in love many years later. The way the two of them looked at each other across the table when the good professor was in his late sixties (and his wife not much younger) made me feel like I was intruding on a romantic evening. In our present context, it raises the question of whether the same result would have occurred had they met online, assuming the possibility had existed at that time. Would their Digital YOUs have "just known" right away?

At the nascent stage of a potential relationship, the Digital YOU usually leaves a giant pile of small artifacts, though rarely is there explicitly sexual or even romantic content on work e-mail accounts or IMs.[3] While *Grazia*, an Italian magazine, recently ranked flirting via e-mail with coworkers (and having sex on the copier) among the worst four things you can do if you are looking for love in the workplace (the remaining two were sending flowers to work and having romantic spats in front of other coworkers), the real reason is likely a combination of the fact that nothing much has actually happened yet and that both parties are hesitant to jump the gun.

People are understandably eager to avoid the humiliation of appearing much more interested in the other person than that person is in them—or ever will be. A wonderful movie example of this occurs in *Something's Gotta Give* (2004), in which Jack Nicholson starts to type an IM to Diane Keaton that simply says, "I miss you." As she waits on pins and needles for his next message, he hesitates. During the long pause, she breaks off the chat session for fear of rejection, leaving him staring wistfully at his screen. This fear of rejection, and possibly of looking ridiculous, is only intensified when the budding romance is among coworkers. Nothing good is likely to come of playing the role of the unsuccessful pining pursuer in the workplace. Plus you still have to deal with the other person in a "business more or less as usual" sort of way if it doesn't work out.

Public social media is a whole other issue, however. There is the understandable tendency to promote a new relationship whether with a coworker or otherwise. Doing so sends an affirming "I'm interested" message to the other person and shows the world that you are out having a life; by contrast silence seems cold and uninviting. The obvious problem is what to do if and when the relationship is over. You can certainly change your relationship status on Facebook, for example—that's why the choice is there—but what you can't do is totally vaporize the relationship's existence. Imagine, for instance, a tool that helpfully tells someone you would like to date that you've dated 29 guys in the last year, with an average relationship length of 15.7 days and 2.1 dates. So if you are concerned about tallying up a long list of relationships that just didn't work out, it's a good idea to follow a consistent policy of avoiding publicizing anything that provides too much detail about your relationships, especially at the start of a new one.

In a work-related context, there's also the issue of whether relationships are considered "appropriate" to pursue. For example, if two people's jobs require them to work closely together, the employer is likely to take a dim view of a potential relationship, since a bad outcome would likely represent the loss of one or the other employee—or both. Coworkers may not be thrilled about the prospect either, if both people are considered essential to the success of their mission. Likewise, if one or both of the potential partners consider their job unusually or uniquely good, they are much more likely to think about whether they should embark on a relationship than they would had they met in a casual social setting. That said, there seems to be little concern about expressing the sentiment that a new colleague is really hot. If someone is physically attractive, it isn't so much stating a meaningful preference or desire as commenting on the obvious while engaging in a bit of same-sex or same-sexual-orientation bonding.

Most people would agree that dating your boss is bad, though we've seen it many times, especially in the case of secretaries. But your liaison doesn't necessarily need to be with your boss or someone else at a higher level in the organization to be seen as a cynical career-acceleration move. While a simplistic view of the world says that the office is broken into clear parts—your boss and those above you, those below you, and those at the same level—the reality is much more complex. Real-world degrees of power and influence cannot be captured accurately by an organizational chart whose full expression is limited to a few levels and pay grades. Thus, you can gain a significant advantage by getting involved with someone who is technically a peer but clearly on the inside track to bigger and better things. In this manner, you gain access to social events, people, and information through digital (or real-world) pillow talk.

When I say, "digital pillow talk," I am referring to the phenomenon of a constant flow of messages of various kinds that become a shared stream of information. It isn't necessarily deliberate, such as forwarding confidential messages or offering up passwords. After all, your computer and your portable device are within the sight and hearing of the other person. So if someone else is repeatedly trying to contact you, both you and your partner will be aware of it. You can't very well shield your screen from someone you're sleeping or living with. Nor can you always mask a highly positive or negative reaction to a message you receive while you're with that person, which naturally prompts the obvious question of what the message was about. In tacit recognition of this reality, colleagues sometimes begin to cc their partner on e-mail that would otherwise be inappropriate, thus altering the contour of the digital grapevine. Whether you do this or not, if one of you is noticeably more junior or less well connected than the other, a new mini-grapevine is likely to form around you. All of this makes sensitive information much less secure than it ever has been in this particular kind of situation.

It may be an unfair advantage for those to whom it applies, but it is part of life. In most offices—at least those that have more than a certain number of people in them— some number of people possess this type of advantage at any given point in time. It means they are likely to find out about promotion opportunities first, get an early heads-up that the company or division is about to be sold, and so on. This is inherently understood by most people, which is part of the reason such relationships and attempts to secure them are usually resented by coworkers. Broadly speaking, the workplace is a community like any other, and it judges its members and their actual or sought-after relationships. Accordingly, commentary by coworkers is one of the sources of evidence we look at in order to identify the existence of relationships between people in the same department or firm.

So what do you do if you are romantically interested in a colleague? Act with extreme caution, especially if you are even arguably the person in a position of greater authority. This is not just because of the possibility of lawsuits, but also because it is very easy for the boss to appear exploitative if the relationship doesn't work out, regardless of the actual reason. Similarly, an underling often may appear to be cynically manipulating an overworked, lonely, and perhaps much older superior. The bottom line is that failed romantic relationships in the workplace are dangerous for both parties, regardless of their relative positions. As a result, the most prudent and probably most often deployed strategy is what I call "digital hanging around."

The idea is simple: if you consider it too dangerous to pursue a romantic relationship overtly at first, you simply observe your quarry as best you can, using whatever means you have available. This makes it easier to form a successful strategy for befriending the person in question, as a step toward the ultimate objective of a romantic relationship, and for determining the best time to make your move. The good news is that observation in cyberspace is easy and won't seem creepy if you're careful. After all, the personal content someone exports on Facebook, Twitter, and so on can hardly be construed as a secret. In fact, if you really follow someone's day-to-day activities for any length of time, the other person is likely to be flattered by the attention. And it puts you in position to be the virtual or real shoulder that person can cry on when something goes awry, which will happen sooner or later.

We can identify digital hanging around by detecting a sustained pattern of multichannel communication that involves many non-work-related events and topics. It is relatively constant, not over the top in intensity, not explicitly romantic, and either unique or at least unusual for the person in terms of the level of involvement in the details of someone else's life. After all, all this monitoring of different media—including mentions of the object of your affection by other people online—so as to be totally up-to-date on every minute aspect of that person's life does take time. Always being on the spot to suggest exactly the right restaurant, movie, or other needed thing does too. So once again, the telltale sign of your strong preference for an individual comes from the amount of time and attention you demonstrably dedicate to that person.

During the course of an actual relationship, little explicit romantic or sexual content is usually generated in the data that can be collected in the workplace. This is especially true in larger companies, less so in small ones. We've speculated that this absence is because "whispering sweet nothings" falls under the heading of things more readily done in speech than in writing and that there are far too many potential pitfalls for all but the most intrepid (or experienced) to attempt it. Something that reads as too smooth may be worth Googling to see whether the writer has borrowed it from somewhere else and/or has used it previously with others.

There may also be a cultural or linguistic dimension to this. Some cultures are more permissive about the degree of romantic demonstrativeness they allow, and some languages lend themselves more readily to romantic speech. For example, contemporary American English lacks common terms of endearment such as *chérie* (roughly, "dear") in French or *tesoro* (roughly, "treasure") in Italian. Such terms offer strong evidence of a romantic relationship, since they are unlikely to be used in any other context.

If and when things start to go awry, one person normally starts giving the other the cold shoulder and becomes increasingly unresponsive, while the rejectee attempts pursuit, sending ever more messages through various media trying to reach out. The exact tacks taken vary, sometimes expressing concern about the other's welfare, complaining about frustration or confusion at being ignored, trying to studiously ignore the obvious by proposing new activities in the face of continued silence, or attempting to pierce that silence with even louder proclamations of love. We know things are headed directly downhill when the messages begin to itemize the number of recent failed communication attempts. We have also seen a number of strategies people use for sending the message that a relationship is over, such as noting how attractive a new acquaintance at the grocery store is, or—good news—they've just happened to meet someone who would be *perfect* for the soon-to-be-ex. Sometimes the person trying to escape sends an interminable list of all of the things he has to take care of before he'll have time to get together again; the list includes a combination of both incredibly trivial items and ones that are impossible to resolve, like ending world hunger. The more subtle forms of rejection often leave their target forlornly peering out of the digital window, hoping against hope to find some affirming tea leaf floating out there in cyberspace.

The form of pining the Digital YOU does is considerably different than what the real you once did in the era when you'd come home at the end of the day and check your answering machine, only to find that you had absolutely no messages. Now, at any time of day or night, there is the IM, e-mail, SMS, blog response, tweet, Facebook wall scribbling, Skype call, and so on that you might get at any moment. Which means that every moment in which there is no sign of life is a letdown. That's a lot of letdown.

It is made even worse in a way that is impossible for pre-digital generations to fully comprehend, since you can now often "see" what the other person's Digital YOU is doing when she's not spending time with you. In the worst case, she's posting and tweeting endlessly about a topic of staggering insignificance rather than communicating with you. The better you know someone's Digital YOU, the more you know where she hangs out online—for example, which online communities she frequents with which identities, whose tweets she has re-tweeted, and so on. All of these are demonstrably more important than being with you. You can spend hours in this fashion, robbed of the more pleasant delusion that the other person was simply too swamped with work to have time for you at this juncture. Such sustained digital pining is also a sure sign that it's over.

At the end of the affair, it is often a no-holds-barred situation, regardless of company size or other factors, probably in part because the dumper knows exactly how the dumpee feels, and others around them have some inkling of what has transpired as well. The facts—or at least the feelings—are already out there. So when the relationship is over, there's often no shortage of explicit and bitter words and, unsurprisingly, one-sided conversations. There may even be some desperate entreaties. A particular example of the latter from one of our investigations is that of a secretary who was having a long-term affair with her boss; in a series of e-mails, she begged him not to marry a different woman whom he had gotten pregnant after a brief relationship, especially since the secretary was now facing a possible divorce because of their affair. (Unbelievably, she discovered his plans to marry the other woman because he asked her, in an e-mail, where to get the marriage license.)

This suggests that it isn't so much a question of respecting appropriate boundaries of behavior as a question of using the most appropriate medium for the task at hand. Happy, romantic messages laden with anticipation are best conveyed by voice. When it's over, however, all the dumpee may be left with is the ability to send messages of different kinds, often with no certainty that they'll be read. The motivation at that point may mostly be venting or an attempt to embarrass the dumper in front of others. If the ending is brutal enough, the fallout is likely to be both public and spectacular. It may include the trashing of the dumper on community sites she frequents; vandalism of personal Facebook and Wikipedia pages, if she has them; or even attempts to engineer high-ranking Google results for her name that will paint an unflattering picture of her. It all depends on how angry and rejected the dumpee feels, how determined he is to seek some form of revenge, how much time he has on his hands, and how Web-savvy he is (or his friends are).

Beware the difference between telling your side of the story and what many would consider cyberstalking or cyberbullying.[4] The online watchdog organization CyberAngels characterizes a true online stalking situation as having the following components: malice, premeditation, repetition, distress, obsession, vendetta, no legitimate purpose, personally directed, [having] disregarded warnings to stop, harassment, and threats.* Leaving aside any potential—and admittedly unlikely—legal complications from such behavior, a revenge-crazed Digital YOU is unattractive at best and downright scary in a Glenn Close/*Fatal Attraction* sort of way at worst. To some extent, whether you achieve this level of scary is a function of what you say, but it is even more a function of how far you go out of your way to say it and for how long after the relationship is over. For example, if you know most of the online communities that your ex frequents, with a little bit of effort, you can probably figure out how to embarrass him in each of them. It is the modern-day equivalent of going to every tavern and coffee shop in town to make sure everyone knows what a complete jerk he is, and it is every bit as "in your face" aggressive. But neither tactic brings the person back to you.

A final important note on the Facebook relationship status setting and any other public social media equivalent: misuse in a fit of pique can be dangerous and even deadly. For example, while it is hard to imagine that most nonsociopaths would actually end a relationship in this public and

---

*This definition is credited to CyberAngels. It now appears at www.wiredsafety.org/cyberstalking_harassment/definition.html and elsewhere.

humiliating manner, a surprising number of people do seem to use it as a means of communicating a difficult message in a way that can't be ignored.

The problem is that if the other person has truly refused to hear the message up until this point, broadcasting it as widely as you can so she'll hear it at the same time as potentially hundreds of others is likely to enrage her—with unpredictable consequences. There have been at least three fairly recent cases in the United Kingdom of a husband killing his wife after she announced her intention to leave him on Facebook. The following are accounts of these cases:

> Wayne Forrester, 34, told police he was devastated that his wife, Emma, also 34, had changed her online profile to "single" days after he had moved out. "She then posted messages on an Internet website telling everyone she had left me and was looking to meet other men," he told police. Forrester drove to her home in Croydon, south London, and stabbed her with a kitchen knife and a meat cleaver.[5]

> Edward Richardson, 41, went to his in-laws' home, where his wife had been living since their separation, after she failed to reply to his text messages following her change in status [to single] on [the] social networking website Facebook. He found [her] in her bedroom and subjected her to a frenzied and brutal attack with a knife and then attempted to take his own life.[6]

> A husband is believed to have murdered his wife before killing himself after she told friends on Facebook they were splitting up, it emerged yesterday.

> Tracey Grinhaff's body was found in a shed in the back
> garden of the family home she shared with her hus-
> band, Gary, and their two young daughters, aged 14
> and 4.[7]

Each of these situations illustrates a different error in judgment, though certainly not one worthy of murder. Perhaps the easiest to understand is the last, where the unwitting husband accidently discovered that his wife was about to leave him; she obviously hadn't left him yet, because she was still living with him when he allegedly returned home to kill her. In the other two cases, the wives had moved on far faster than had their soon-to-have-been-ex-husbands. Whether it was a question of just a few days or a few weeks, it was simply too soon for these men to see their wives actively seeking new male companionship in such a public fashion. It probably isn't a coincidence that in the Richardson case, in which some weeks had passed since the wife had moved out, the estranged husband had tried (in vain) to contact his wife directly in various ways before his alleged murderous attack, whereas in the Forrester case, according to news reports, the injured husband went straight for the cocaine—and then, apparently, the meat cleaver.

It is easy to dismiss these as isolated instances of crazy people. But even normal people can snap if severely provoked. Grinhaff, for example, was an electrical engineer, a professional with absolutely no history of aberrant behavior. Neighbors actually described the couple and their children as the "perfect family." It is always important to remember the old adage about there being a thin line between love and hate, and even more important to keep in mind that the line is a whole lot thinner when someone's ego is publicly on the line. Consider that people who are dumped via a tweet, a relationship status change, and so forth are left no wiggle room, no way to save face. They can't credibly claim to have dumped the other person, especially not if they were recently texting happily away about their couplehood. Worse still, they may not even be able to claim to have had the remotest clue that things were going south. That they are likely to lash out in response to such an unwelcome event shouldn't be a surprise. There is no substitute for direct dialogue, whether online or in person, especially when the news is hard to take. It may not be easy, but as the preceding examples show, it beats the alternative.

If the notion of your soon-to-be-former partner coming at you with a dangerous kitchen implement seems too unlikely to be motivating, and you aren't the sort to care about good karma, consider that there are many digital equivalents of the meat cleaver. For example, a masterfully written blog about how you mercilessly broke someone's heart can do real damage to you for a long time. This is especially true if you have the ill fortune to have a relatively uncommon name. But even if yours is Jane Smith, someone savvy will be sure to add enough appropriately placed keywords, such as the name of your current employer, to identify you specifically.

Note: Laws to protect you from slander by scorned lovers—or anyone else you happen to have pissed off—are still nascent and currently seem to require a combination of repeated, deliberate acts of online stalking; possible threats of real-world violence; and some horrible result or damage, such as attempted suicide. However, one thing seems certain: neither our legal system nor law enforcement is even remotely equipped to deal with massive numbers of such claims. In addition, to qualify as slander, a public statement has to be demonstrably false. It is hard to imagine that the law will ever provide protection against statements that are totally true just because they are unpleasant or unflattering, no matter how widely or artistically broadcast.

If one change to the ecosystem in the next few years is more scorned lovers learning Internet-style revenge, another is the ongoing de-stigmatization of the use of online dating sites, dating people you first met online, and online nudity. None of this is really surprising. The difference between using a social media site to say you're available and participating in an online dating site is not a big one. If people today spend much of their waking hours online, they will increasingly be dating people they meet in this fashion, with the result that the first few weeks or even months of the romance will occur in the digital world. Likewise, the increasing level of casualness with which people post pictures online—of people at parties or bars at 3 A.M., when they've first woken up, or at other non-Kodak moments—continually blurs the moving line that separates acceptable taste from voyeurism.

There is little that teenagers who have grown up with MySpace, Facebook, Flickr, YouTube, and the like aren't comfortable doing or putting online. This, and the prevalence of cell phones equipped with cameras, has led to the surprisingly widespread phenomenon of *sexting*, in which teenagers send partially or fully naked pictures of themselves to others from their cell phones. I suppose doing this may mistakenly seem like a safer, more controlled, yet still sexy means of foreplay. It has become the equivalent of what flashing once was. Unfortunately though, sexted pictures aren't gone in the blink of an eye and can cause all kinds of problems for the sender and recipient(s). Bewildered law enforcement authorities are left to determine whether to prosecute anyone who distributes the pictures if they are of a minor, including the minors who took the picture of themselves.

The problem of breaking the law is only one reason that sexting is just a dumb idea. If the picture is originally sent to more than one or two people, it will likely find its way to many more, including total strangers, drawing attention from a much wider audience than a relatively small group of horny teenagers. Further, as Internet search engines become more and more focused on multimedia and on more intelligent disambiguation of people, it is not at all implausible that teens who sext now will be confronted in a few years with a college admissions officer who searched for their name online and found one of the sexted pictures in the automatically selected profile picture that came up. This is not only plausible, but probable, since most search engines would—not unreasonably—consider the "best" picture of someone to be the one that is in widest circulation. Still, none of this is likely to stop the behavior once a large number of people have become accustomed to it in their formative years. Indeed, as today's sexting teenagers turn into adults, sexy pictures are likely to continue to be an important marketing tool in their online dating efforts as they compete for the optimal mate in the increasingly global dating market.

Though online dating services are now ubiquitous and pretty much universally accepted, I think the question of whether they actually make the prospects for love better for those seeking it in cyberspace remains open. Certainly these services allow you to connect with an otherwise impossibly large number of people who have been pre-filtered by your specified criteria. In fact, many of these services seem to actively encourage consideration of as many different dimensions of compatibility as possible. Clearly they understand how to do marketing. The real world offers no way to specify a large number of parameters and end up with a list of (theoretically) possible partners. Dating sites do.

But in so doing, they promote the idea that you really can find someone who is compatible with your desires in every way, that an instant spark will ensue, and that anything less is settling for less than you deserve. And who wants that? There is, however, a problem, which is best illustrated by a simple exercise: take all the men or women (as appropriate) you have ever met and, using their different traits, construct your perfect mate. For example, you can decide that you like one man's gentleness, another's sense of humor, yet another's physical appearance, the way a fourth one made you feel, and so on.

If you actually take a few minutes to ponder this truthfully, you will likely find that you have selected attributes from several or more different people. And that's normal, because (1) no one is perfect in every possible respect, and (2) no one is totally perfect by your particular yardstick. Even when a dating site serves up a long list of theoretical possibilities, no one candidate from the list will be the best in every respect—even when only measured against the others on the list.

This emphasis on optimizing across many dimensions of possible goodness naturally ups the ante. Someone who wants to appear either more like you or more likable *to* you can do so easily enough, given how much information there is about most of us online. The temptation to do so is strong, as it is both easy and likely to succeed—for a while, anyway. As long as your relationship stays in the digital world, where so many answers are readily available with a few clicks and keystrokes, it's great.

For example, anyone who wishes can quickly find out various things about me, including that I love dogs, where I went to school, my various philosophical views on managing and running a business, and so on. So it is not rocket science to appear compatible with me in many respects, at least initially. Someone really obsessive could try to see what I've already said on any given subject before responding to any particular message I send, with the result that someone who might at first appear to be a soul mate is actually a search-engine-aided mimic who types really fast. But this person would be unmasked quickly in a first real-world date when there would be no way to look up the "right" answer.

This may sound over the top, but it isn't. If anything, it is just being practical. Digital courtship has become a remarkably competitive thing; the more members sign up for online dating services, the longer the list of potential matches for nearly everyone. It's arguably not that different from sending your résumé to a company that receives many such résumés. Not all candidates are going to get to the interview stage; indeed, most won't. Since no one likes to be rejected, using strategies to try to make sure you are more likable and more closely attuned to the potential date's wavelength than are other candidates arguably makes sense. Assuming, of course, that you do nothing to contradict your established public persona, lest the other person is also diligently doing research on you.

This kind of situation is one of the reasons that tools that paint detailed pictures of people's online personas are likely to become commonplace in the next few years. For example, while people aren't likely to have blogged about dogs unless they have a dog, there is little risk that someone has a well-established record of dog-hating on the Internet. By contrast, someone's political opinions are likely to be discoverable these days, since it has become so common for people to comment online about their views on elections. Thus, a sudden change in party affiliation to increase your compatibility with someone, while perhaps flattering, might well be counterproductive.

The more dimensions that seem to align, the greater the pressure for an in-person meeting. If you've already met once or twice in real life (or have met online via some kind of video), but the vast majority of your interaction is still digital, there's some natural difficulty. After exchanging hundreds or potentially even thousands of messages within a fairly short time, how exactly should you behave when you are confronted with the actual person, whose mannerisms are still unfamiliar and whose sense of what stage your real-world relationship is at may differ considerably from your own?

If finding, or even cementing, true love online is hard, satisfying sheer lust is much easier. For one thing, the number of search dimensions is radically reduced, because things like shared hobbies, long-term lifestyle choices, and so on simply don't matter. Many dating websites exist only to satisfy specific fetishes; indeed, this is a fairly common usage of YouTube as well. Then, of course, there are the professionals.

A couple of years ago at Cataphora, we began working on the infamous D.C. Madam case. The case centered around a woman named Deborah Jeane Palfrey, who had for some years run a high-class call girl ring in the Washington, D.C., area. Many of her clients were powerful people, and she was known to be discreet. At least until she was prosecuted for a variety of crimes years after closing up shop. U.S. Director of Foreign Assistance Randall Tobias and David Vitter, a senator from Louisiana, were identified as clients; many even higher-profile people were alleged to have been. What interested us about the case was that all of the surviving evidence was phone records. This was, in large part, because the events in question mostly occurred years ago, before Twitter and Skype, at a time when most people still thought that e-mail accounts were much like snail-mail addresses, rather than being instantly disposable without leaving much of a trace.

The nature of the argument we were asked to make was as follows: Ms. Palfrey indicated that clients patronized her business to have their fantasies fulfilled, but not to have sexual intercourse per se. Her clients were not men who had difficulty finding willing and attractive sexual partners; rather, they turned to her service to fulfill their unconventional desires in a discreet manner. In other words, she was marketing and selling fantasies; if the girls consented to sex without her knowledge to get a better tip, so be it. In a meeting with us, she famously noted that the laws with respect to prostitution require sexual intercourse to occur, but "there's no law against a woman getting on all fours and barking if someone wants to pay her well to do so." Specifically, she and her then-attorney argued that such detailed fantasies took a noticeable amount of time when clients called to book appointments—much more than would a simple instruction such as "Please send a blonde to the bar in the hotel on the corner of Fifth and Broadway at 10 P.M."

So the idea was to subpoena phone records—not just hers, but those of similar services, as well as straight prostitution rings—in order to show that there was a substantive difference in the call patterns. (In the end, we never got to do it due to events, including Ms. Palfrey having to change lawyers due to a jurisdictional technicality and her subsequent suicide.)

Certainly, many other Deborah Jeane Palfreys are still out there, and I suspect that many of their clients now avail themselves of everything that is available online to help them better and more efficiently detail their fantasies. After all, if a picture is worth a thousand words, how many is a video worth? I doubt that people call such services and spell out URLs, but rather use some kind of anonymous online identity exclusively for this purpose. In fact, in a totally unrelated investigation, we once had a target who had more than 200 e-mail identities, one per woman with whom he engaged in conversation on different S&M sites. While this was doubtless done as a precautionary measure to shield his real-world identity, the 200 digital identities all had many similarities that strongly suggested they all belonged to the same person.

In addition to the obvious fact that they all existed in the limited context of S&M-related communities, they all unsurprisingly made the same spelling errors, tended to surface at about the same time of day, and always exhibited the same preferences. The right computer program could easily determine, with a high degree of probability, that all 200 were the same guy. Even without a subpoena or other brute force method to determine someone's real-world identity, aggregating enough data across public websites might well unmask him. For example, perhaps he posted something on a movie-related site, enthusiastically reviewing a mainstream film that in some way related to S&M, with the same spelling errors, grammatical quirks, and so on, and he posted it under an identity that was more easily linked to his real-world name. After all, the more identities, user names, and so forth that someone has across an ever-growing number of online communities, the harder it is to keep track of them all. Over time, for any number of reasons, a person's level of squeamishness or concern for personal privacy over a particular type of content in a particular forum may change, making it even harder still.

There's going to be big money in de-anonymization software, or software that links up otherwise anonymous online commentary and other behavior with specific people. This is a way for advertisers to get data they really want but are currently prevented from getting either for legal reasons or by their own privacy policies. Like all situations in which there is a big golden carrot for those who get out there and do it first, many people will try to come up with this software. As always, not all of them will do an equally good job. Only it will be much worse in this situation than in most, because de-anonymization is a really hard problem when performed at the scale in question.

The result is likely to be terrifying to a great many people for quite a long time, because sometimes the wrong data will be associated with the wrong person. But rather than a collection of individual results on Google that, at least in principle, everyone understands may correspond to different real-world people (who obviously live in different places, are in different professions, and so on), what will be delivered is one aggregate profile of you. Or at least the person the software thinks is you.

The fact that the information may be presented with some probability of correctness will hardly blunt the problem, as few are even likely to notice it. Further, as noted in the example of the two Matt Welshes in Chapter 2, sometimes it is next to impossible for a person or machine to guess correctly. Let's say you happen to share a name and one or two other incidental similarities with someone who is a next-generation client of someone like Ms. Palfrey. Combined search engine and de-anonymization software would helpfully tell anyone looking for you that your list of hobbies and interests includes travel, skiing, wine tasting, and that you like to have _____ done to your _____ while you're _____. It will quickly make you nostalgic for the days when your biggest worry was a credit rating that somehow got mangled through the wrong identity. That's a real nuisance, but not one that causes mysterious smirks and whispers whenever you enter a room for months to come.

In the final analysis, despite the many respects in which mating rituals are different for the Digital YOU, one important thing remains the same as ever: some amount of mystery is essential. Like giving too many details about your prior romantic escapades when you first start to date someone, a totally predictable pattern of behavior—like tracking someone's every move—wasn't attractive 100 years ago and won't be attractive 100 years in the future either.

When the Spotlight Shines on You

AT CATAPHORA, WE often hear concerns about privacy issues in the workplace, but people are usually afraid of the wrong things for the wrong reasons. Here's what you should fear: Imagine total strangers spending weeks, or even months, going through every shred of electronic content you have created or amassed in the past three years. Now imagine they are all lawyers. If they find anything embarrassing about you or that paints you in a less-than-flattering light, they set it aside in a special folder by category (such as "Jane considering cheating on her husband" and "Jane questioning the competence of the head of manufacturing") for their bosses to inspect. When they're done, they question you about everything they've found for several hours or several days. They may repeat this exercise multiple times. And if you lie, getting fired is probably the least of your worries.

If you think that sounds bad, wait—it gets worse. These lawyers are theoretically on *your* side, or at least on your employer's. Once a good chunk of the data they've collected from you and about you—usually about 20 percent of the data in a legal dispute—is handed over as evidence to these lawyers to actually review, the lawyers on the *other* side will get their crack at it. Some plaintiffs' firms pay rich bounties for any evidential finds that help their case. Sometimes an especially juicy find that makes a big case will literally yield a new car that shows up on the doorstep with a big bow on top as a thank-you—not just in movies like *Erin Brockovich*, but also in real life. Such plaintiff attorneys, whose business model depends on winning rather than billing hours to their client, are especially motivated treasure hunters. Part of that motivation is to make you and your employer look as bad as possible, since they'd like to avoid a sympathetic opponent.

If you are the CEO or another top executive, the lawyers on your side will scrutinize every data record to limit the amount of your data that has to be turned over to the other side. But this is an expensive proposition, not to mention a lot of work and attention to detail.

The good news is that if you happen to be a bit lower down on the food chain, "your" lawyers probably won't spend all that much time going through your data with a fine-tooth comb. After all, your demise probably wouldn't shave even a nickel off the stock price. And proof that you knew about something isn't nearly as damaging as proof that someone three levels above you did. The bad news is that potentially damaging or embarrassing documents of yours will still slip through and be handed over to the lawyers for the other side. So if you've committed any electronic indiscretions related to the matter at hand, you may have cost your employer the case and yourself a job.

Obviously the scenario that I'm describing is a lawsuit. According to a recent study, approximately one-fourth of all U.S. companies had employees' e-mail subpoenaed in a lawsuit within the last 12 months.[1] That number is actually strikingly high, since the vast majority of firms in the United States aren't megacompanies that get sued all of the time, but rather small businesses with only dozens or at most hundreds of employees. What most people don't stop to consider is that any kind of major stupidity committed by both corporations *and* individual employees often ends in litigation, involving what almost everyone would consider an intolerable invasion of personal privacy.

The distinction I'm making here is best explained by way of example. Companies may conspire to artificially inflate profits or fix prices, but individuals solicit kickbacks. Companies may get hit with class action suits for gender-based discrimination; individuals proposition their assistants. Regardless, in all of these instances, the corporation will be sued. The cold, hard reality is that the longer privacy is protected in an adverse situation, the greater casualty it eventually becomes. Worse, fairness often becomes an even greater casualty.

This hits home much more readily in smaller businesses. For example, at a company of Cataphora's size, which is roughly 100 people, everyone clearly understands that the time and money that goes into defending the firm against a lawsuit comes out of their own pockets. In the event of a money loss that isn't covered by insurance, innocent people will likely lose their jobs, because there's only so much money to go around. That kind of realization necessarily shifts their perceptions of reality.

You are probably thinking that while parts of this may be horrifying, it is nothing that will ever happen to you, as opposed to your boss's boss's boss.

Think again. In many types of litigation, the net is cast much wider than you might think. Often the intention of a subpoena is to gather data from anyone who *might* have been in a position to observe or know certain things, such as potential safety issues with a product, a manager sexually harassing his subordinates, the frequent violation of a particular policy, and so forth. Consider for a moment how many people this might be—or more to the point, *who* they might be. Anyone in an office is in a position to observe harassment or to have been a target of it, from the college hire to the most senior people in the company. Similarly, many people might be in a position to hear rumors about or otherwise deduce a problem in product testing or manufacturing. A good lawyer knows that at least some of this knowledge—or even speculation—is likely to be memorialized somewhere in e-mails and IMs. At any rate, it is certainly worth looking.

Junior employees are usually less savvy about how comments they perceive as normal griping could quickly turn ugly in the event of a lawsuit. So even though teams of lawyers are less likely to swarm over their data, warm guns—if not actual smoking ones—are likelier to be found here than in the data of senior executives who probably have not one, but several, lawyers on their speed dial.

To take a simple example, complaints about management are common in many companies. While many people will avoid blasting their direct manager, at least in writing, higher-ups constitute more public figures. In other words, they are considered fair game, in much the way celebrities are. In reality, much of this flaming is totally meaningless because it is uninformed; for instance, a junior employee opining that the CEO is a jerk because the company's stock price went down. But let's take a few examples in which the commentator appears more credible:

- "Bob doesn't understand what is in half the reports I send him."
- "Sally is so hands-off she wouldn't notice if sales in Asia dropped off by a third."
- "Do you know who they ended up hiring? I bet she's tall and blonde."
- "Ray bragged that he could convince the investors of just about anything. He said that half of them were senile."

All of these comments suggest some direct knowledge of events within the company. They might seem like relatively harmless quips or nothing outside the ordinary. But here's how they become problems for their hapless authors in the event of a lawsuit: The first two suggest negligence, which is always a good basis for a lawsuit if real harm to anyone arguably resulted from it. The third suggests at least two different types of hiring discrimination; the incompetence of managers suggested by the first two comments could also be used in an argument for discriminatory hiring practices, if other circumstances supported it. The fourth comment borders on fraud; deliberately lying to investors is a big no-no.

Such quips unavoidably become evidence when they appear with "clearly responsive" content; in other words, in a document that clearly has to do with the topic of a subpoena, such as a defective product. The lawyers for your employer's opponent would cackle with glee at any reference to a negligent or incompetent manager whose job function related to the product in question. This is how your passing, long-since-forgotten, wiseass jab can become the basis for the "real" reason your company's product killed someone.

To say that you can forget about that year-end bonus would be a gross understatement.

Many things become relevant when everyone is under the microscope. In one case we worked on several years ago, a number of middle managers at a large company (unsurprisingly) claimed they were innocent of a scheme to inflate revenue. When interviewed by the attorneys, all of them indicated that, although they had hoped to do better than they did, they were relatively satisfied with the performance of their business unit, given the conditions of the market and so on. In other words, maybe they weren't collectively turning in a stellar performance, but it was passable and certainly not worth breaking the law to make it look better.

This was, in actuality, a plausible argument given the basic facts of the case. However, the argument was quickly undercut by our staff. More than one of the managers was looking to jump ship, a sure sign that something was about to go very wrong—or already had.

One of the things we have learned over time, and hence have baked into our software, is to scan for people who are updating their résumés, sending them out, or inquiring about other job opportunities. Anyone is entitled to leave his job or seek a better one (though not on company time). But people rarely try to leave jobs in which they are happy or that they think are going reasonably well. Most people don't like looking for employment. It takes time and often involves facing a certain amount of indifference or rejection. Plus, if word gets out, it could hurt you at your current job. Keeping all this in mind, if there is evidence that several people are starting to look for other jobs, it is pretty clear that the forward outlook for the department or company is not a rosy one.

It doesn't take a rocket scientist to figure out that it might not be a good idea to look for your next job using the laptop from your current one. But few people consider this particular ramification. Generally, they have no clue how just about any of their actions become a focus of attention if they are really under the microscope for some reason and all of their data is being examined by their employer and the employer's attorneys. Here are some other examples we've seen in e-mails:

♦ Fessing up to a relationship or personal financial problem to a friend could be used as evidence that you were distracted from your job; had you not been, the bad thing the other side is alleging would not have occurred. It could also raise concerns that you might be willing to divulge critical company secrets for badly needed cash.

♦ Online receipts for certain types of items in your mailbox offer a fertile ground for treasure hunters. For example, if you ordered a book on a highly arcane type of security, you will now be hard-pressed to claim that you were a novice with respect to anything more than simple stocks. In one case we worked on, we found receipts for Nazi memorabilia in the inboxes of some of the key executives. While totally irrelevant to the matter at hand, when disclosed, this discovery understandably made the defendants less sympathetic to witnesses. In other words, it increased the probability that former colleagues would be willing to rat them out.

Just as we've noted that you can protect your privacy to a greater degree by using Gmail or other private accounts, you can also help protect it by avoiding mixing sensitive personal content with business content. As mentioned in Chapter 4, we call e-mails and other types of data records that mix the two *commingled communications*. These are inherently dangerous, as the following true story illustrates.

A company was sued for fraud relating to its charges on a large and complex contract. Numerous internal communications indicated genuine confusion on what the actual billing rules were. We like to see such evidence of confusion when we are working with the defense, because the law requires that there be provable intent for fraud. Being in a hopeless muddle, as these people clearly were, is the opposite of having a clear intent. Since there is more cluelessness in the world than outright dishonesty, this is not an uncommon situation. Where the case took a turn for the unusual was in its many commingled communications.

For example, our system automatically joined an e-mail that mentioned a "spike in the billing in the month of March due to . . ." with a different and unrelated exchange of messages that contained a totally different type of reference to spikes, specifically the bondage and S&M variety. The system joined them together because some of the bondage-related messages also made passing reference to the contract in question.

This created the possibility that the plaintiff company might be able to "out" the executives engaged in the behavior to advance its interests. That there was nothing actually illegal about the behavior doesn't mean it would not have been embarrassing if publicly exposed. And especially for publicly traded companies, avoiding such embarrassment can easily be worth settling an unrelated lawsuit in a way that is favorable to the opposing side.

While such exotic examples illustrate the point that you should never mix personal and professional text, far more mundane behaviors are starting to cause big problems, especially in the financial world. We've noted elsewhere that contradicting yourself about how good a particular product actually is can get you—and your employer—in really hot water. So if you have ever gotten up on the wrong side of the bed and decided that whatever you do for a living just sucks, I have one important piece of advice for you: keep it to yourself. Or at least never put it in any kind of written form.

One of the most common types of problematic communications is that of the personal opinion that would have been best left out of any electronic media anywhere but that gets tacked onto the sort of content that is certain to be subpoenaed in the event of a lawsuit. Ironically, the more true the opinion is, the more damaging it is. Consider that if your opinion is that the CEO is a space alien from the planet Zylox, it is not likely to be given much weight. However, if your opinion is that, based on your observations, she has a cocaine problem that led to the tanking of a critical deal, that would be a different matter. Hypothesizing in writing that your boss is a cokehead is probably unwise under the best of circumstances. Doing it while at work, on your employer's computer, is even more foolish. However, juxtaposing that hypothesis in the context of something specific, such as a particular named deal or product, makes the e-mail, IM, or other communication subject to subpoena in the event of a lawsuit involving the collapse of the deal, a defect in the product, and so on. In other words, it becomes the type of cosmically monumental stupidity that stories will be told about for years to come.

In some companies, expressions of personal opinion that, at best, would be severely frowned upon by the powers that be are so commonplace that most people probably don't give them a second thought. A good example of this occurred in September 2009, in a case brought against the large banking firm UBS. In discussing a particular kind of subprime-mortgage-related security known as CDOs that the firm held, a UBS employee wanting to unload the securities in the days following rumors of an impending downgrade sent a colleague the following e-mail: "OK, still have this vomit?" He wasn't trying to ask if the recipient had recovered from stomach flu. You are probably wondering what the big deal is. After all, everyone makes mistakes, and all this guy was doing was acknowledging such a mistake. Surely there's nothing wrong with that.

The difficulty was that UBS was still hawking CDOs to its customers at the time. This gave Connecticut Superior Court Judge John F. Blawie an opportunity to display his sense of humor:

> The court takes UBS employees at their word when they referenced their Notes, these purported "investment-grade" securities which they sold, as "crap" and "vomit," for UBS alone possessed the knowledge of what their product, their inventory, was truly worth.[2]

Part of the problem in this instance was that the derogatory terms were apparently in widespread use relating to these securities. Thus UBS's attorneys could not argue that the e-mail's author was an isolated naysayer. (Judge Blawie ordered UBS to set aside $35.5 million to cover a potential judgment against it in the case.)

To properly put all of this in some sort of context, it is helpful to know at least a little bit about the process behind a lawsuit. The following provides a brief and simplified synopsis.

The first signal of litigation that is visible outside the legal department is often what is known as a "freeze letter." This is a letter—usually in the form of an e-mail these days—from the legal department that informs you that you may not delete any data relating to certain topics. Or in extreme cases, any data at all. (The IT department will generally have backups of everyone's data, but the intervals at which backups are done may vary.) Freeze letters also can serve as a means of telling you that you may not delete any relevant future messages or documents you receive until told otherwise.

If a subpoena is actually issued and you are named as a "custodian" or relevant person, someone will usually come and collect your data. This generally gives you an opportunity to at least exclude data that is clearly irrelevant. Sometimes, however, your data can be collected without your knowledge, either automatically by IT or by companies that contract to provide such services surreptitiously. Such companies will send in armies of people in janitorial uniforms who won't take out your trash but will leave with copies of your computer's hard drive. Then, as described earlier, some or all of your data will likely be reviewed by attorneys, or perhaps paralegals. Potentially all of your data will be analyzed by a system such as ours, since human time is expensive, but computer time is essentially free (once the computers and necessary software licenses have been purchased).

If for some reason you turn out to be particularly important in the dispute, you will probably be deposed under oath by the attorneys for both sides. You might even have to testify in court, though the vast majority of lawsuits never actually get that far. Content that you have produced or seen may be used as part of this process, either to help refresh your memory or to help make sure you will tell the truth. Visualizations, such as the graphs and matrices in this book, may also be used.

You can easily be particularly important without having a fancy job title; assistants or secretaries often fall into this category. They have direct access to key actors in the litigation and may also have close personal relationships with them. Many see all of their boss's e-mails and documents. Often the best defense from the corporation's point of view is to try to pin any potential problem on the lowest guy on the totem pole so as to limit its own liability. The point is that this really can happen to almost anyone. All that matters is the situation and your relationship to it.

What you need to know:[3]

♦ Once you receive a freeze letter, under no circumstances should you delete any data that could even arguably be covered by that letter. There is a high probability of getting caught, whether by us or someone else, and the consequences could be severe. If you get caught telling your friends to delete relevant data, it will be even worse. If you get some really dodgy e-mail—such as one from a friend and colleague saying that he really needs to stop doing Ecstasy before important client meetings—the best thing to do is simply to delete it immediately, before it can be backed up (unless of course a freeze letter requires that you preserve it). There's also nothing wrong with suggesting to others that they do the same. It is only a problem if you know that litigation requiring that item to be pre-served is already in the offing.

◆ Don't wait for a freeze letter to apply common sense and realize that a situation might turn ugly. One example of this is the increasingly disgruntled employee. While at the beginning you may be supportive of some of her grievances, the train may go off the rails at some point, leading to either the person being fired or quitting, and then suing the company. And lo and behold, messages you had long since forgotten will now form the basis of the plaintiff's case.

◆ It is important for you to try to understand what the nature of the case is. If there is anything in your e-mail, IMs, or other documents that is going to be bad for the arguments your company is trying to make, it is gener-ally best to get out in front of it and bring the content to your manager's attention. Hopefully, the lawyers will say, "Oh, that's nothing," and that will be the end of it. But even if it's not, generally the amount of trouble you will get into is directly proportional to the amount of trouble your employer has because of you.

◆ Similarly, if you are a central character in the drama and there is anything in your data that may simply make you look bad, even if it's unrelated to the case, you may want to get out in front of it. Managers don't like to find out from attorneys that you have a prescription drug addic-tion or that someone is trying to extort money from you.

◆ Always clearly mark any communication with an attor-ney as "ATTORNEY-CLIENT PRIVILEGED COMMU-NICATION." This means it does not need to be disclosed to the other side. However, it is important to note that the titles of documents (or the subject lines in e-mails) generally still must be disclosed, because documents that are being held back for reasons of privilege still have to be logged in what is called a *privilege log*. This gives the other side an opportunity to challenge the validity of the privilege claims.

Although there are many totally legitimate reasons for employers to pay attention to the actions of their employees, I started this chapter by talking about litigation for two reasons: (1) it is the equivalent of a Digital YOU cavity search; and (2) sadly, many unchecked problems often end up in litigation (or a credible threat of it), at least in the United States.

Even though corporations carry insurance that pays for many types of lawsuits, they generally try to avoid litigation (unless it is a case from which they stand to profit.) Even if the insurer foots the bill, it is costly in terms of time and constitutes an unwanted distraction. Juries can behave in entirely unpredictable ways, completely ignoring the judge's explanation of the applicable laws. And there is always the possibility that facts exposed in one lawsuit can end up triggering other, larger lawsuits.

At various points throughout this book, I've asserted that companies really just aren't that interested in the day-to-day doings of their employees as opposed to the positive results they produce and any problems they might cause. Simply put, companies care very much about the profit and loss potential associated with each employee. Sure, individuals within a corporation might be motivated to spy on one another for various reasons, but companies frown severely on this kind of behavior. It doesn't make for a healthy working environment, and it is likely to end in a lawsuit against the company.

However, this self-interest doesn't mean the same corporations aren't obliged, for an increasing number of reasons, to monitor their employees' communications. I have heard that as many as 70 percent of U.S. corporations monitor all of their employee communication. *Monitor* in this context means that a software program scans all messages and documents looking for a specified set of things that are determined to be bad. This set generally includes the following:

◆ Distribution of porn, especially kiddie and other illegal porn
◆ Evidence of other illegal activity (such as someone dealing cocaine from his office)
◆ Use of racial epithets and other derogatory terms that imply discrimination
◆ Violation of some specific company policy designed to prevent fraud, theft, or legal liability
◆ The leaking of company trade or other secrets to the outside world, especially to competitors and the press

In other words, anything that is manifestly and directly harmful to the corporation raises a flag. Anything the computer finds will be reviewed by a human being to determine whether it is what it appears to be. False positives (for example, distribution of a newspaper article concerning an illegal activity), which are frequent, can be dispensed with quickly and don't amount to much of an invasion of privacy.

Where there appears to be smoke or actual flame, an internal investigation will be opened. All of this costs the company both time and money, but for many businesses, it is a far better alternative than doing nothing at all. It is a simple matter of self-protection, and not just at the abstract level of the corporation, but at the personal level as well. The reality is that if you are a manager, you are likely to be held responsible for what occurs on your watch. Yet you can't see the e-mail with the sexist joke or the IM arranging some kind of offbeat erotic encounter unless it is sent to you or you have some kind of monitoring software in place. The most terrifying aspect of this from a management perspective is that the offending e-mail or IM may have literally taken a minute or less to type. The author could be hardworking and productive, and even have enough sense not to send such communications to obviously inappropriate people. In other words, bad things happen very quickly and with resounding silence.

If you are wondering what the real harm is in such an isolated event, let me provide a few examples:

- ◆ A sexist joke can become a real problem in the context of a class action suit involving discriminatory practices against women.
- ◆ Every e-mail program's helpful auto-completion of names inevitably trips up the best of us at one time or another—and sooner or later, it will send the wrong message to the wrong person.
- ◆ If a clandestine tryst is between coworkers and the romance breaks off, the message that suggested it could be used as evidence of sexual harassment or coercion.
- ◆ As noted in the example about the financial spikes versus S&M-related spikes, if a practice is kinky

enough and either of the participants is an executive, it could be used to embarrass the person—and hence the company—in the event of completely unrelated litigation.

If the people in question had conducted these exchanges via their Gmail accounts, they would have been much safer. Yes, it is still possible to accidentally send a message to the wrong place. But overall, by using Gmail instead of your corporate e-mail to send out personal messages, you are greatly increasing the probability that what happens in Vegas stays in Vegas—perhaps even literally.

The truth is that the passive monitoring arrangement used in most companies seems to bother relatively few people, since they assume they have neither pedophiles nor racists in their midst. And if one should be hiding out undetected, they have no problems with that person being unmasked. Further, a computer program reading your messages doesn't elicit the same kind of emotional response as an actual living, breathing entity doing so.

While most people don't think of it this way, the ubiquity of e-mail, IMs, and other forms of digital interaction puts employers in a real bind with respect to many kinds of liability by concealing behavior that was once readily apparent. This means that more and more employers are going to spend more and more resources monitoring their employees.

This is best explained by way of example. In my first job after college (which I'm sad to say was before e-mail was common), there was a guy who would obviously ogle all the women in the office. He would hang around their cubicles and keep a keen eye out for any of his targets going to get

a cup of coffee or to the cafeteria so as to have the opportunity to engage her in conversation. While mostly people
just gently chided him or cracked jokes about it, his manager
eventually had to have The Talk with him that enough was
enough.

The interesting question is what this same personality
would do today. Would he still physically eye and stalk his
prey to the same degree in the physical world, in full view of
the entire office, or would he continuously IM them and so
avoid the humiliation of prying or mocking eyes? Would he
track them in the real world to see how they looked each day,
or would he settle for Facebook updates? If the latter, how
would his manager know there was a potential issue, short
of a complaint being made?

Unfortunately, waiting for a formal complaint to be made
is a bit like locking the proverbial barn door after the horse
is already in the next county. Imagine a lawyer for someone
who feels she has been sexually harassed trotting out a giant
box of paper in court and holding up one piece after another
to show the jury just how many not-really-necessary or just
plain inappropriate messages the plaintiff received from the
defendant in a relatively short period of time. Then, as on
TV, the lawyer would turn to the jury during his closing
argument, point at the box, and say, "Do you really believe
the company didn't know this was going on?"

Most people's knee-jerk reaction would be to agree with
this argument. On the surface, the accusation of the company's culpability is compelling enough that the existence of the
box of printed communications probably causes more lawsuits to be filed. But it is, nevertheless, a spurious argument.

The natural and most convenient way to respond to an
IM that you feel is inappropriate is with another IM rather

than, for example, with a phone call that someone else might overhear. Moreover, the pursuer needn't even be in the same building as his target, so there may have been no opportunity for a manager or an HR person to observe uncomfortable interactions between these two people in the hallway or the coffee room. Here's where it gets really interesting: the in-person interactions may be markedly different from the digital ones.

To see why this might be the case, I'll use myself as a simple example. When I am logged into my company's internal IM system, I get lots of IMs, many of which are just inappropriate distractions from my point of view. So I don't log in much, because I know that any important would-be IMs will manifest instead as phone calls or e-mails; in my case, most people in the company could simply come into my office and interrupt me.

Most don't though—even with our smallish size. While the exact reasons may vary, the basic one is always the same: people who think nothing of taking someone's time and attention with digital communication often hesitate to so much as cross the threshold of the office of someone they respect or admire—or that of their boss's boss. Perhaps they are having a bad hair or bad skin day; perhaps they dribbled toothpaste on their shirt that morning or have socks that don't match; or perhaps they get cosmically tongue-tied in potentially stressful face-to-face situations, whether it is with an important prospective client or an attractive new colleague.

If, despite these arguments, you are still thinking the poor schnook deserves his privacy unless he really upsets someone, ask yourself the following question: why should today's schnook be entitled to more privacy than the same schnook a generation ago?

This is an interesting question, and not relevant only to schnooks. Consider the following:

♦ Old-fashioned mail rooms made no claim whatsoever to enforcing any degree of security or privacy. In fact, the manila mail envelopes were often secured only by a string tie and generally had holes in them. (I presume the holes were there to prevent empty envelopes from accidentally being circulated. If so, it is a clear indication of what the relative priorities of efficiency versus privacy were.)

♦ As noted in other chapters, both in-person conversations and phone calls were likely to be at least partially over-heard by others. Individual offices with thick walls that block most sounds are expensive, so most people don't have them. The fact that e-mail, IMs, Twitter, and so on have removed sound from the equation, and therefore provide apparent privacy, is a side effect rather than the intention of these media.

♦ More time spent on various forms of electronic commu-nication generally means less time spent in face-to-face meetings. Yet when I think back to my days in larger companies, such meetings were how I actually got to know most of my colleagues—except for those sitting in the same row of cubicles. Industrial psychologists have performed many studies showing that even relatively small changes in a physical office floor plan really change behavior, including increasing or decreasing the produc-tivity of a team. Simply put, even before e-mail became ubiquitous and the average fitness level of office work-ers sank even further, walking a few rows over to talk to someone or see if he wanted to go to lunch was a pub-licly visible statement of personal preference that everyone could see. The Digital YOU knows no such physical boundaries. It's possible to interact with anyone, at any time, with considerable privacy, not counting the presence of monitoring software.

Here's another highly counterintuitive thought: from the standpoint of liability reduction, the people who are most worth watching are the higher-ups (middle management and above), not the little guy. The reasons are obvious if you stop to think about it:

- The higher you are in the food chain, the greater access you have to the company's most valuable resources, such as proprietary information, cash, and important customers.
- An ogling schnook in the mail room may, under the wrong circumstances, cause some legal issues, but they are nothing compared to what a horny executive can do.
- Lower-level employees simply lack the opportunity to do things like make illegal agreements or decide to cut manufacturing costs in a way that might pose safety hazards.
- A bad manager has a far greater ability to generate misery for others—and hence misery and problems for the company—than does an individual worker.

Spending lots of time to (possibly) catch otherwise-productive, lower-level employees committing minor infractions isn't a good investment. Even if there are many thousands of such employees, the cost of determining and pursuing each infraction exceeds the cost of the infraction itself. This isn't to say that some companies won't do it anyway. But in most firms, it isn't good business judgment.

That said, it is important to understand that in order to catch serious problems, *all* communication has to be filtered through risk management software, because there are plenty of bad things that literally anyone can do. No one wants to find out that the cops have carted away one of their employees for a serious offense, or that a low-level employee socially engineered her way to all kinds of company secrets that leaked out and caused the company's stock price to plummet temporarily.

While the software may "watch" everything, human beings must ultimately answer the question of what types of things should be flagged as worthy of investigation or, indeed, worth bothering about at all. This may sound simple, but in reality it is a fairly nuanced process. For example, most people would agree that drug dealing on the company's premises is totally unacceptable and nullifies any considerations of employee privacy. But what about an employee with a close family member who has a drug problem that the employee is discussing with various counselors? Most software would have difficulty differentiating the former case from the latter. So the simple rule of thumb is to avoid discussing anything that even borders on topics for which the company would have a legitimate reason to want to set up content filters.

Some companies have far greater risks relating to litigation and compliance than others. Mostly this has to do with the industry sector in which they operate. For example, pharmaceutical and biotech companies are almost always knee-deep in both lawsuits and regulation issues, because while their products save lives, in some instances, they also kill people. By contrast, more traditional businesses may rarely see even small lawsuits and have no regulation concerns. An attorney I once met, who worked for a large retail clothing chain, joked that the most harm his company's products could do was to make people look fatter than they actually were.

However, there are many kinds of business risk that merit employers' keen interest, many of which have nothing to do with any kind of legal issue. These tend to be much more equally distributed among industry sectors.

Here is a very basic one. Companies usually trust the opinions of their line managers when it comes to managing and evaluating employee performance, even when it gets contentious. It is a bit like when a cop gives you a ticket for running a red light. You may deny it. You may not have actually done it. But the cop has all the credibility in this situation, and you have none. After all, you don't want to pay for the ticket, and the cop was just doing his job.

Problems with managers are therefore usually diagnosed after something bad has already happened, such as several of their higher-performing employees resigning. Even assuming there is nothing more sinister going on than a manager having proved the Peter Principle true and there being better opportunities elsewhere, several high-value employees all leaving the same department can be crippling, even to a large company—they have important knowledge that no one else currently does. Ironically, by voting with their feet, they are helping to resolve the situation by bringing it into clearer focus. The far more insidious situation is one in which employees stay put despite poor management but are dangerously disengaged to the point of making mistake after mistake.

Good HR departments often have a gut sense of when they are dealing with a weak manager. But they are often understaffed and even more often lack the power to cause changes in the organizational structure, at least until there has been clearly visible damage. Not surprisingly, the people reporting to the weak manager are often afraid to speak up. As a result, some astonishingly bad managers are left in place for years, being displaced only when something *really* gets screwed up. In the meantime, they can profoundly damage the lives of the individuals in their department.

It isn't that most corporations don't want to hire good managers or rid themselves of bad ones. But most managers are neither stunningly good nor so obviously horrible that they should be axed. And for all the reasons already noted, many of the real-world cues on which managers and HR departments traditionally relied now can only be seen in some roughly equivalent fashion in the digital world.

For example, it was once far easier to see how much a manager directly interacted with the members of her group. Back in the day, a manager staying ensconced in her office most of the time was considered a warning sign in many companies. The acronym MBWA (management by walking around) was once a catchphrase for good management style. It meant that the manager was expected to proactively and frequently interact with his employees in person, not just to ensure that the employees were receiving proper direction, but with the aim of building personal rapport. What replaces MBWA today is an interesting question, especially since so much digital communication is inherently not one-on-one interaction.

It is therefore understandable that many upper managers are more than a little panicked about how they are going to cope with a world in which they can directly observe so little of what goes on and in which they have less control than ever before over the flow of information in the "echo chamber" (see Chapter 2). That said, because so much business communication is now memorialized electronically, there are many never-before-possible checks and balances on managers:

◆ Weak managers tend to avoid providing direct negative feedback in employee reviews or other communications, because such feedback is likely to provoke confrontation and possibly cause offense. Yet most employees aren't perfect in every way, and providing negative feedback is a necessary and important part of a manager's job.

◆ As noted in earlier chapters, we can measure how evenhandedly a manager interacts with his employees. Less professional managers clearly have their favorites and broadcast these preferences, demoralizing everyone else.

◆ We can also measure how much the manager—or anyone else—demonstrably exerts leadership. A weak manager will often fan the flames of resentment that occur in response to bad news such as wage freezes, layoffs, or other unpopular announcements; a strong manager will quell them.

Further, as we've seen, managers' assessments of their employees can be given a sanity check via automated analysis of electronic communications and other data:

◆ How often someone is passed the digital buck. Insecure managers sometimes hesitate to admit how much they lean on underlings in this respect.

◆ How often others consult an employee for her domain expertise and how wide-ranging her sphere of influence is. In some companies, such important contributors are overlooked at raise time, because their impact is distributed across the many people they have helped in the course of the year.

◆ How many bits of distinct work product are in widespread use for any individual—for example, a really good slide, a clever sound bite, or a clear explanation of something complicated—can provide an objective metric of how much practical intellectual capital an employee is contributing to the company.

Indeed, these kinds of measurement tools can be directly helpful to the manager herself. It is all too easy for managers to lose track of what all their employees are doing or what impact they are having, especially if these managers have many people reporting to them.

In the final analysis, the overall result of so much business communication being recorded and the use of the type of technology described in this book is a far greater degree of transparency rather than a one-way mirror. This isn't to say that technology won't be misused. However, technology itself neither fires people nor sues them—nor does it make value judgments about behavior as opposed to statistical ones. It merely provides an objective tool to view things as they are, not as you—or others—think they are or wish them to be. And what better use of technology could there be?

The Twists and Turns of the Digital Grapevine

COMMUNITIES OF DIGITAL YOUs don't expand, transmit information, or otherwise behave in ways that conform to our real-world experience. Perhaps the choice of nomenclature is partly to blame. The term *community* has traditionally implied something that is firmly rooted, in which the participants have a significant investment, such as a physical neighborhood, a church group, or a professional group of specialized practitioners. Yet the whole guiding principle behind many online communities or groups is the ease of connecting—in other words, the opposite of requiring a considered investment of time, effort, or money. As we'll see, although there are reasons for this, it can sometimes backfire, and in memorable ways.

It is the kind of fantastic tale that could only occur in Silicon Valley, one of arrogance, initial meteoric success, semi-random chance, and, above all, millions of dollars quickly spent and quickly lost.[1] It is a story that has more than a little to do with the topic of this book, since it involves the company that most would argue was the first high-profile social networking site. No, not Facebook or MySpace, but a company called Friendster that began about three years before MySpace was sold to Newscorp for half a billion dollars.

By all accounts, Friendster was plagued by the expected problems of a Silicon Valley "it" company; with an initial $13 million in venture capital and press galore, but far too little accountability, with problems ranging from poor business decisions to engineering scalability woes to management infighting. Then something unexpected and really bad happened.

In early 2004, Chris Lunt, who was the director of engineering, noticed an odd phenomenon: Friendster's peak activity level was at 2 A.M. local time. That was unusual, so he decided to look into it. When he looked at the IP addresses of the people accessing the site, it quickly became apparent that despite the problems already mentioned, the company had achieved its goal of national market dominance. The only problem was that the nation in question was not the United States, but the Philippines.

Chris soon unraveled what had led to this unexpected state of affairs: a friend of Friendster's CEO named Carmen Leilani de Jesus was among the original set of Friendster members—specifically, number 91. Carmen was at the time a San Francisco–based marketing consultant, hypnotist, and sometime dominatrix of Philippine origin. She knew a great many people, not surprisingly including many in the Philippines. The Philippines is not a wealthy or especially technologically sophisticated country. However, one thing for which it is widely known is the extraordinarily high percentage of its citizens who work outside their native country[2] and, thus, far from their family and friends. The Philippines was an ideal market for Friendster, though the site was arguably functioning more as a widely used convenience than a community. It continued to become even more ideal as Filipinos became an ever larger chunk of Friendster's user base, the inevitable consequence of an increasing number of Filipinos flocking to Friendster just as an increasing number of Americans—and others—were discovering MySpace and Facebook. Unfortunately for Friendster, this was an exceedingly undesirable turn of events. Big American brands were unlikely to spend their big American advertising budgets on a largely Filipino customer base. Further, there was no good way to reverse the course of events. Sure, Friendster could have blocked access to their site from IP addresses in the Philippines, but Filipinos were all over the world, including in the United States. The company therefore couldn't block them. And even if there were some way to identify and then discriminate against Filipino users solely on the basis of their ethnicity, it would be a PR nightmare at best and a legal and civil rights debacle at worst. This turn of events spelled the end of the Silicon Valley dream for Friendster. The company did, however, continue to flourish in Southeast Asia, which accounted for more than 90 percent of its 115 million users by geography and probably closer to 100 percent counting expatriates. In December 2009, it was finally acquired by a Malaysian company named MOL.[3] To this day, the Philippines remains Friendster's top market.

What is not at all unusual about the Friendster story is the often curious and unexpected nature of the digital community. It is a big mistake to think of it as being analogous to real-world coffee klatches, bowling buddies, sorority sisters, or the like. For one thing, such groups are often fairly homogenous. In the real world, if you happen to enjoy both tractor pulls and poetry readings, you are unlikely to invite a friend from one of these circles to an event in the other—at least not without careful consideration as to the probable consequences. But when launching an online community, the owners promote it aggressively to everyone they know even slightly; both Grandma and Carmen, their former bosses and the people they just started dating. They would never invite all of these people to the same real-world party, but they think nothing of inviting all the Digital YOUs to their community.

Here's an even bigger difference between digital and reality: you have to actively and deliberately choose to participate in real-world groups. It requires you to exercise a preference, to express a choice. You can't be at both the ballpark and a classical music concert at the same time. Indeed, you may not have time to do both things in the same day or even in the same week. If your free time is in short supply, you may have to prioritize quite ruthlessly. As a result, the number of real friends, people in whose day-to-day lives you can participate in a truly interactive relationship, is limited. Note that you can use the Social You-niverse view in the Digital Mirror software to see how people, represented as planets orbiting around you, drift closer to or farther away from you over the course of time (see Figure 8.1).

Although you only have bandwidth for so many real friends, you can have just about an infinite number of acquaintances. Some people consider the term *acquaintance* to be a dig, meaning that a person is somehow not significant enough to have "friend" status yet—or possibly ever. Others use it to indicate that they aren't willing to vouch for someone by calling him a friend. Often this is simply because they don't know the person very well and don't want to be held responsible if he turns out to be a real jerk.

The online world has unfortunately done away with the notion of acquaintanceship. Everyone is, increasingly, a

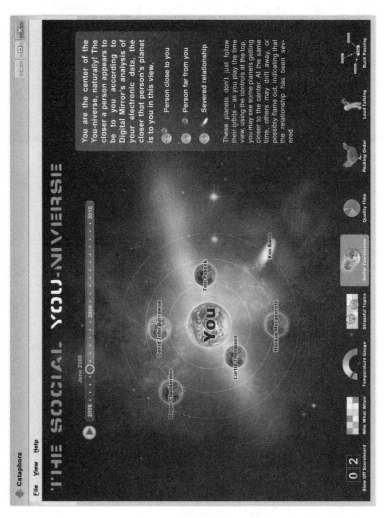

Figure 8.1 Social You-niverse view (*Background photo courtesy of NASA*)

friend of everyone else. It's free, and more important, it feels good. You are more popular now than you ever have been, more popular than your parents ever dreamed of being, even if they were the homecoming king and queen. And since "unfriending" events are relatively rare, you are certain to get even more popular over time. Who wouldn't want that?

The problem that Friendster and all of the social networking sites that followed it have is that the social network they are trying to facilitate is actually based on (at best) acquaintanceship and not on friendship. As a result, its shape and character are much more difficult to predict. It made sense that Carmen networked with her clients, of which she apparently had many. After all, the cost in time was minimal, the cost in dollars zero, and it was likely good for business. It further made sense that a free, globally accessible online community like Friendster would quickly become popular in a far-flung ethnic community for which it provided a taste of home.

Despite the fact that there are almost certainly more dominatrices per resident in San Francisco than there are in most other places, Friendster was at least somewhat unlucky. For example, instead of Carmen, Friendster CEO Jonathan Abrams might have known a leader in the gay and lesbian movement, a New Age guru with a large following, or any number of other people who belonged to groups that have ample numbers in the Bay Area. This would have resulted in a customer base with real economic buying power, and Friendster might now be a household name instead of a bizarre footnote in Silicon Valley lore.[4]

The problem might also have been averted had Friendster imposed a charge for use—specifically, a connection fee for each new friend. Certainly that would have considerably slowed the site's growth. But it also would likely have made the type of audience growth more predictable and more controllable. Consider that as an individual, I am far less likely to "friend" 500 people if the cost is $5 a head. Thus any one person is much less likely to be able to influence the shape of the community at a gross level once that community gets very large. Likewise, far fewer people would ask me to friend them for fear of being rejected over the $5. (Paradoxically, the lower the cost of friending someone, the greater the probability of it occurring, but the greater the humiliation of rejection.) However, such "put your money where your mouse is" connections would be far more meaningful and real. The cost would simulate the few dollars you spend in gas, parking, bridge tolls, or subways to hang out and stay connected with your friends in the real world.

To the best of my knowledge, no online communities currently support the notion of a "really good friend." It is easy and effectively risk free to make distinctions between close friends and friends, or between friends and acquaintances, in private conversation. Just think what you would say if your sister or best friend were thinking about dating someone you know and asked your opinion. You would be unlikely to go out on what might be a limb. However, the public notion of different tiers of friends immediately creates a number of difficulties that almost everyone seems to want to avoid. Stop for a moment to consider what the impact of any of the following would be:

◆ You can only have 10 really good friends on your favorite social networking site at any one time. You can change them as you wish, but you can never have more than 10. Alternately, after the first 10, the cost

for each additional really good friend is $10 more
than the prior one.

◆ Friending "really good friends" costs $20 per person.
Regular friends are $5. Acquaintances are a really
good deal, at only $1 a head.

◆ If you want to recommend someone on a site like
LinkedIn, it will cost you—or her—some real money.
And you can only highly recommend five people at
any one time.

◆ A "quality time" tax rather than a financial one
means that friends with whom you don't interact
enough over a long period automatically get demoted
to acquaintance status and then, at some point, to
former acquaintances.

Now further imagine that all rejections, demotions,
and instances of unfriending one person to make room for
someone else were logged and people could pay to see them.
Imagine what consternation that would cause, as people
tried to weigh the value of saving $5 today versus irritat-
ing someone who could come back to haunt them later. The
point is that the way people communicate and who they
passively allow into their communities or networks is much
more heavily affected by business models than by anything
real or true about relationships.

Any online business model that too closely simulates
real-world situations is a dangerous thing. It is, after all, far
more pleasant for users to have the illusion of 306 quality
professional contacts and 192 friends than the reality of a
handful or two of each. It is one of the reasons the Digital
YOU is way more cool than the real you. Not to mention
the detail that both advertisers and investors require online

communities to have lots of users to stay in business. And imposing a cost would immediately pose a barrier to unlimited use. In the end, though, fear of humiliation would come to represent a far larger barrier still.

If the Digital YOU is less afraid of public rejection than the real you is, that is totally understandable. After all, those doing the rejecting may be thousands of miles away and, in many instances, are only rejecting one of the Digital YOU's many easily shed personas. But perhaps it has more to with the fact that the Digital YOU rarely ever has to taste explicit rejection and naturally has less of an aversion to it. For example, if actual dates don't happen all that often between online "friends," much of the rejection can be handled passively and relatively painlessly; if there's no response for a long enough period of time, most people will get the message. Plus, no one is keeping track of foxylady09 or hotsurferdude22 deciding to pass on you. By contrast, someone who works or lives down the hall from you in the real world would likely be forced into an explicit "no" much sooner. Word of it would spread, the story of the rejection getting embellished a bit with each retelling, thus upping the pain level of the rejection. And there's no easy way to just withdraw from your environment if it gets uncomfortable. Likewise, if you post your résumé somewhere online, chances are a certain number of employers have perused it and rejected you without you having so much as an inkling.

The flip side of this is that the Digital YOU's first encounters with real rejection are likely to sting all the more because of the novelty of the experience. Imagine a scenario in which all but three of your several hundred friends evaporated, even if the cost of keeping you as a friend had just become one lowly dollar a year. Poof, just like that, and for all the

world to see. Imagine the orgy of Digital YOU voyeurism
on such a theoretical Judgment Day + 1, as millions of Digital YOUs rushed to see who among them still retained the
lion's share of their friends and to gawk at those in (hopefully) worse shape than they were. While this type of event is
probably still a ways off—and certainly won't happen overnight—the era of promiscuous friending will likely draw to
a close at some point. Here's why:

The longer any of these popular networking sites are
in existence, the more interconnected everyone in them
becomes. In the end, if there is no way to discern the meaningful links from the "I wanted to be polite" ones, these
sites become increasingly ineffective—at least for most useful activities. For example, someone trying to connect to
me seeking a job will often stress that she knows someone
I know. But more frequently these days, I have absolutely
no idea who the person in question is, despite the fact that
he was at the same place at the same time as I was, perhaps
many years ago. It doesn't bother me much, but it probably
bothers the job seeker immensely.

This is a thorny problem with no easy resolution. Change
is scary, but staying the same leads to inevitable obsoles-
cence and irrelevance. My advice: don't get too attached to
the idea of your Digital YOU being surreally more popular
than the real you. It is unlikely to last forever. And when it
comes to an end, you don't want to have nurtured a needy
Digital YOU that sacrifices its dignity by love bombing or
stalking former friends in an attempt to reconnect at almost
any price.

It is surely no coincidence that one important class of
online community is immune to the real-world dynamics
of dollars and the ego associated with others' perceptions.

These are the gaming and fantasy sites in which people shed their real-world identities and are completely disconnected from their daily lives. In these magical, hermetically sealed worlds, they can be anyone; friend, unfriend, or insult anyone; or, for that matter, shoot or plunge swords through anyone they choose. All without the danger of losing a good job opportunity or other undesirable real-world side effects. A whole book could be written just about these massively multiplayer online role-playing games (MMORPGs), in which people can participate in a totally different type of complex society with its own rules, loyalties, markets, histories, and individual communities. In short, these games offer risk-free, immediately gratifying mass entertainment at a relatively low cost. (These sites often charge their users for all kinds of privileges and accessories. Users seem willing to accept these charges to be able to play the games; after all, they are not being asked to pay to participate in real life.)

Unfortunately for corporations, information flows through the digital grapevine as freely as friendship does. In the United States, this dynamic is aggravated by the fact that people change employers with much more frequency than they did a generation ago. As a result, an increasing number of people view their "career equity" as being vested in their relationships within their industry rather than in a given employer. This in turn means that for many, the most valued community in which to have standing and influence is one that cuts across the corporate firewall. The millions of dollars that large corporations spend on securing sensitive corporate data can't secure their most volatile data resources: their employees.

While the leaking of sensitive information is a real problem, the frequent dispensing of fairly mundane tidbits about what it is going on internally is arguably becoming an even bigger problem for many companies. Most employees don't have access to board-level secrets or anything close. But almost all employees hear rumors of layoffs or impending benefit reductions, or notice that it has been a while since any of the higher-ups have had a smile on their face. Anyone can diagnose a morale problem, observe that the fresh flowers have disappeared from the lobby and been replaced by plastic ones, or noticed that the coffee has suddenly disappeared in favor of odd-tasting brown water. By talking about such slice-of-life nuggets online, employees release information about the company to the public.

The business of sites like glassdoor.com is to encourage both employees and interviewees to anonymously discuss what they think of employers. Personal impressions, petty grievances, entreaties to upper management, and various other things are aired on such public sites in increasing numbers. Even if such activity is counter to some corporate policy—such as disclosing salary information is in many companies—employers are usually unlikely to pursue individuals. The amount of hassle and expense to accurately establish a specific person's identity (if it can be done at all) is potentially large, whereas the damage caused is likely minor. That said, depending on the size of the company and any given division of it, it may be quite easy to unmask the employee's identity. To see why, consider that the combination of specific complaints, overall level of business or technical knowledge, correctness of grammar and spelling, and the apparent level of the person's seniority with the company are just some of the factors that can be used to quickly whittle down the possibilities—perhaps to just one or two people. In this event, there's no telling whether a manager will act out in a fit of pique against a perceived culprit who has violated a corporate policy and/or made him look bad. I think there is little doubt that such sites skew toward the naturally negative. All marketing people know that an angry customer has at least ten times the impact and visibility of a happy one. It's the same with employees, and arguably even more so, since the likelihood is that more ex-employees comment than current ones.[5] It may be unfair, but that's life. Employers do have a countermeasure, though: they can have their own marketing and human resources people anonymously post positive reviews.

None of this would fool a competent investigator or HR attorney. To borrow from Tolstoy, all happy employees in a company are happy in more or less the same way, but all unhappy employees are unhappy in their own unique way. Reviews from contented employees, whether real or concocted, will up the company's rating, and that may influence many job seekers. But it doesn't erase the combination of specific individual details and overall patterns of those who were unhappy. For example, angry people tend to remember details about an incident that angered them, even fixate on those details. Things like how many people were laid off, how many were hired overseas, how many were promoted above them, or when exactly an incident occurred are often woven into the narrative. By contrast, happy employees tend to talk more about their overall feelings than about specifics.

The glassdoor.coms of the world *do* offer a useful perspective for the investigator or litigator. For example, do the complaints seem to align with those typical of other employees at the same company? Are there common distinct complaints, praises, or other comments associated with the company? If so, are similar, not terribly common phrases (such as "frustrating and opaque") being used by multiple people, suggesting either active collaboration or at least a single strong influencer? An individual complaint that is generally consistent with known issues about a company is more likely to be true than is one that seems to come from out of the blue. However, a complaint that is identical in various details to an existing one may be a copycat, someone looking to cash in on a perceived gravy train, such as a victorious lawsuit.

From the point of view of the job seeker, the key is to look at what the commonly voiced complaints seem to be and determine how seriously to take them. Complaints that are too similarly worded may well have been written by the same disgruntled person—just as similarly glowing reviews may well have been written by the same HR person. The specific issue may simply be irrelevant to your job function. It is important to keep in mind that one man's poison is another man's pleasure: what some employees might really like about a company, you might dislike intensely, and vice versa. Lastly, always remember that such forums tend strongly toward the negative. In both the real world and the digital one, genuinely contented people usually don't spend much of their time and energy telling people how happy they are about this or that; they are spending their time doing whatever it is that makes them happy.

Online financial forums such as Yahoo! Finance function in a similar way to such sites, despite various differences in their usage models and the fact that their primary target audience is investors rather than job seekers. Ultimately though, any socially acceptable online forum that provides at least the illusion of anonymity will bring both boosters and detractors, real and artificial (but mostly insiders, either way) out of the woodwork. Of course, you needn't have a forum to anonymously give commentary about a company; something like Twitter will do just fine. All of this forms a complex picture, but not a totally unmanageable one. Software could be written to capture the equivalent of the Digital YOU for a company.[6] However, whether such software would meaningfully change behavior or just create the need for a new kind of PR job is unclear.

By contrast, controlling employee access to specific corporate data is not an especially hard trick. The difficulty is that employees must know what they need to know in order to do their jobs well, and once a person knows something, it is the *knowledge* that must be contained rather than the *data*. That is a much harder problem. *Data*, by definition, refers to a particular set of words or symbols occurring in a particular order. Software can easily identify copies of specific data—or even near or partial copies. However, knowledge can be expressed in a wide variety of ways. For example, someone leaking sensitive information would likely omit specific details that only a few people would know, while still conveying essential material that a much wider set of people could have surmised. Or the person might distribute nuggets of information across different messages, online forums, and/or anonymous personas.

While there have always been a variety of motivations for leaking information or otherwise discussing internal issues in a public forum, historically most have related to disgruntled or embittered employees. Revenge or spite isn't the only motive. Sometimes it is as simple as the satisfaction of having thousands of people follow your comments when your own managers won't give you the time of day— something that is far likelier to occur when you are dishing juicy insider morsels. Or there may be the idealistic hope that anonymous employee comments on popular sites may actually bring about real-world change.

Recently though, there has been a new and different motivation. A well-placed employee—especially a highly observant one—is often in a position to know many things well before investors, other employees, press, and the guy on the street do. Discretely sharing some of these insights can create otherwise impossible opportunities and career shortcuts. A couple of years of always being right gives you a certain mystique and sense of omniscience, two very desirable attributes to have when you are looking for that next job—and a larger paycheck.

The key, of course, is that people don't do this under their own identity, because it would get them fired in a hurry. But online identities are both easily had and easily abandoned. If a particular identity acquires a significant following, its real-world owner will at some point step forward to reap the rewards. If such an identity is leveraged to get a job, the new employer may try to co-opt it for their own purposes, such as using the identity to say favorable things about a forthcoming product. The employer may not own the voice of the identity but does control the paycheck of the associated human. But then the game merely begins again with new identities. To make this system sustainable, friends employed at different companies share such information among themselves—for example, Gina at company A tipping off Greg at company B about an impending large event at company A, so Greg can tweet or post about it. This way, everyone seems smarter, and it looks less like anyone is parceling off the inside dirt from his or her own company. This technique only adds to the mystique and makes it harder for any investigator who might try to put the pieces together.

This is an interesting new type of what I call a *shadow network*, one that is not easy to detect because its motive is neither monetary theft nor revenge. Its purpose is merely to make someone look clever, to build a reputation for a set of online identities that have no manifest connection to any one employee—or, indeed, to anyone at all. If those involved are smart, they won't tread anywhere near disclosing insider-trading-level information. But what is insidious about this from a corporation's point of view is that quite a bit of damage can be done even without any laws being broken. Just as with individuals, many small pieces of data pieced together—none of which are individually protected—can paint a clear and compelling picture of reality. For example, let's take a large technology company where many middle managers openly participate in a "dead pool," in which people place bets on which promising employees are going to resign next and digitally high-five when their choices win. Something like that is not confidential company data or a trade secret, nor is it a material event such as a looming mergers and acquisitions transaction. But if something like this is happening, it is an important and unambiguously bad sign. If news of it gets out and can be substantiated, it can surely impact the corporation's stock price.

Of course, if employees aren't smart, the more direct gains to be had from insider trading are likely to appeal. Shadow networks, such as alumni groups or people who all worked on the same presidential campaign, can be spread out over many companies. The information trading may often start off with other motivations, such as the desire to be seen as clever. But with each hop the information takes in the network, the likelihood increases that someone will purchase—or dump—securities in a way that will light up the radar at the Securities and Exchange Commission. Whole careers in the financial sector have been made this way and unraveled accordingly. The case of Raj Rajaratnam of Galleon, who is said to have made tens of millions of dollars in insider-trading profit in the tech industry, is one of the best known. According to the *San Jose Mercury News*, the paper of record in Silicon Valley:

> They were Raj Rajaratnam's friends, college classmates, business associates and former employees, but federal prosecutors say they doubled as the hedge fund billionaire's secret network of Silicon Valley insiders.[7]

Other articles noted the difficulty that prosecutors had in sorting through the different, sometimes barely interlocking, circles surrounding him. It was certainly not a question of one coordinated group of people; rather, it was smaller trusted sets who exchanged information or gossip among themselves. Indeed, some were likely people from whom information of value might be gotten someday. What they all had in common was that they had more loyalty to Rajaratnam—or at least their particular small circle of friends— than they did to their Silicon Valley employers such as Google, Intel, and AMD.

Employers are scrambling to put together policies to try to protect themselves against their employees ever commenting on them in public at all without permission. But the law isn't necessarily on the company's side when it comes to employees' activities conducted outside the office from their own computers. Nor do employers have any way of enforcing such policies, because proving the link between an anonymous online identity and a specific human being is usually not a simple task; it requires the identity to have had significant activity, so that the Digital YOU behavior patterns between the anonymous identity and suspected employees can be compared. Given all of this, most employees would only thumb their noses at such policies, rendering them less than useless.

Thus, my advice to managers and business owners is as follows: Don't spend an excessive amount of time crafting such a policy. Content yourself with issuing a high-level statement, such as saying that any unauthorized public statements about the company that violate any aspect of its general policies could lead to termination. Using commonsense examples, explain to your employees how foolish public commentary on the company can boomerang back at them. Most important of all, remember that the fewer real problems there are, the fewer will be reported (or invented) by anonymous people online.

9

Can the Digital YOU Improve the Real You?

SO MANY OF us are creatures of habit. We have a routine for everything we do in both the digital world and the real one. Barring some significant life change, our routines get more and more fixed over time, because they represent a coping mechanism—a means of navigating through the many small hazards and annoyances that surround us daily. After a while, we don't even think about how we get things done anymore. Life simply is what it is.

While this is perfectly understandable, it is also a form of stagnation. We aren't trying to improve anything about ourselves or those around us; we are merely travelers on the path of least resistance. As more and more aspects of our lives are projected and recorded online, that path can increasingly be visualized down to the most minute details, such as the fact that we work longer hours toward the end of the quarter, have a predictable increase in communication with friends as basketball season starts, or regularly blow off some things—and people—for days after or just let them fall through the cracks.

Though our actual behavior may change little over time, our perception of it changes considerably. Nostalgia is a

powerful distorter of past reality. It blurs and fades bad
memories to a far greater degree than it does good ones,
convincing many of us that the job, relationship, car, house,
or life in general we had at some point in the past was incom-
parably superior to what we have at present.

A great example of this is the friend who, at about the
six-month mark in a new job, starts to complain about it.
The complaints reach a crescendo, and around the one-year
mark, he is again polishing his résumé. The quick sandwich
down the street at lunch gives way to a one- and then a two-
martini lunch. His behavior is predictable insofar as it usually takes a few months in a new job to learn enough about it
to see where the problems are; within a year, he'll likely have
figured out that at least some of these problems aren't fix-
able. Maybe he is even right. However, since no job will ever
be problem free, his behavior is a pattern of certain failure.
Yet it is not an uncommon one.

Here's why. Especially by the one-year anniversary, the
pain level associated with problems from the job immediately prior to this one—and the job before that—will have
started to recede into distant memory. So, by definition, the
current job is both (1) worse than the prior job, and (2) less
attractive than it was during the initial honeymoon period
when he was acquiring 10 new friends a day—at least according to Facebook. However, since the exact problems vary
from job to job, he fails to look in the mirror and realize the
obvious repetitive pattern.

The Digital YOU has a potent weapon against such
reality-bending nostalgia. With tools such as the Digital
Mirror, you can literally see the inevitable mood souring that
takes place within a week or two of your six-month anniversary. And you can know that within another three to four

months, short of a highly unlikely act of God, your attitude needle will be well into the red zone. Most important, however, a quick glimpse in the Digital Mirror will remove your nostalgic beautification of prior jobs. You can cross-reference past irritants that provoked frustration and anger by time, topic, people involved, and your level of reaction at the time. If the prior job actually made you angrier at the time than the current one does, you will quickly be reminded of it.

If confronted with such inescapable hard facts, including your own documented sentiments, would you find some way to avoid digging the same hole for yourself again and again? Or would you only become more depressed because the chances of finding the ideal job seem so slim?

If you are someone who has dated quite a few people, the Digital YOU offers a basis for keeping score. How long was that last relationship really, and how long did it take before it started to go south? Was it before or after you introduced the person to your family or to your good-looking housemate? But more than just keeping score, you can predict outcomes and perhaps even causes. You picked the movie? Bad news—you have only a 22 percent chance of another date. The restaurant? Even worse—only a 14 percent chance. (If you talked extensively about your ex, no calculation of probabilities will be needed.)

While most people are willing to alter such surface-level choices if they see a clear benefit to doing so, changing ingrained reactions is an entirely different matter. For example, if the person you are interested in becomes the pursuer, sending noticeably more messages over more channels to you than you do to her or tries to "friend" all of your friends, do you start seeing Glenn Close in *Fatal Attraction* and run for the hills?

If you could see such clear patterns in your relationships, would it significantly change your behavior? Or would you still believe that when you meet the "right one" it is magic, not models, that dictates?

As a longtime manager, I have often seen a manager express surprise when one of her employees requests a transfer—or just decides to leave the company altogether—despite the fact that none of the other managers can ever remember her having anything good to say about the person. Usually the manager is totally unaware of it and will even vehemently deny it. What she doesn't realize is that many people are managers precisely because they continually focus on identifying and resolving problems. As such, when they talk with their colleagues, they most often discuss problems. Perversely, better employees are often tasked with wrangling down the really hairy problems, some of which may have no satisfactory solutions, and therefore may get fewer pats on the back than their more average coworkers who reliably complete their much simpler tasks. Simply put, good, consistent performance is often taken for granted, but the odd failures are complained about.

If managers were shown unambiguously that they were consistently critical of the employees on whom they relied to perform the most difficult tasks and that 99 percent of their comments about these people over time were negative, would they change their behavior?

At one time or another, we've all had something bad happen to us. Whether it is a death in the family, a failed relationship, or a lost job, it is a hard enough blow that friends feel the need to offer support. Software can pick up such events relatively easily by identifying a combination of denser-than-usual communication with friends or family members and the presence of negative sentiments. However, as the old saying goes, "A friend in need is a friend indeed," and sadly not all friends do rise to the occasion. In an era where just about everyone seems to have more friends and professional relationships than they could ever have time for, it is seductively easy to believe that you have lots of friends who will take you out for that post-pink-slip drinking binge or "there's other fish in the sea" ice cream. But that doesn't make it so, and if anything, no one may do anything at all, because there are demonstrably 602 other people who might.

Do you rise to the occasion when something bad happens to a friend who was there for you during a crisis? Do you reciprocate even if you are snowed under with work or are on vacation? Do you even clearly remember whether different people you know helped you during one of your bad times? Would your behavior change if angel-on-your-shoulder software helpfully prompted you when you got an SOS message from someone who once helped you?

Imagine if your sister told you that you generally take about a week to return her messages. With the notable exception of those occasions when you need her to cat-sit. Your first reaction would likely be "That's not true. You're exaggerating. When I'm not really busy, I call you back right away." But then she provides the digital proof—quickly and easily with a few mouse clicks. As a result, you learn that it is actually more like two weeks than one, because the response times are much shorter when they involve cat-sitting, so those messages should not be part of the average. For a few minutes, you'll properly feel like a jerk. But what happens the next time your sister reaches out? Or the next time you need a cat-sitter?

Each of the preceding examples illustrates plausible real-world scenarios of how we may be confronted with hard data that suggest clear ways to improve ourselves and have more control over what goes on around us. But even with so much compelling data sliced, diced, and visualized in software, can years of built-up, often subconscious behavior be changed in a flash?

The answer, at least in most cases, is of course not. In the end, the Digital YOU is nothing more than a projection of the real you in cyberspace, a version of yourself that has been freed from certain real-world restrictions and thus arguably more fully reflects your true personality. Personalities may change somewhat over time, but usually only in response to traumatic life events or sheer necessity.

So let's not talk about redesigning anyone's personality. But what if you really could make yourself a better listener? Make other people hear you more? And suppose you could do this just by observing your own behavior on an ongoing basis, noting what seemed to have a positive effect or a negative one—for example, that less loud talking causes your digital voice to carry further. Would it really make any difference? Would it change your life? Or is it already too late, since those around you have already adapted to your various quirks?

Ultimately, there's only one way to find out. So although the first version of the free Digital Mirror software we are making available with this book necessarily performs only a very small subset of what our full investigative software does, we hope you will use it to get a glimpse of the Digital YOU. What you do from there is entirely up to you.

Notes

Chapter 1

1. The courts may decide that this is less true when it comes to mobile devices that are paid for by the employer but are naturally used outside of the office as well.

Chapter 2

1. Charles Naquin (DePaul University), Terri Kurtzberg (Rutgers University), and Liuba Belkin (Lehigh University), "Being Honest Online: The Finer Points of Lying in Online Ultimatum Bargaining," presented at the Annual Meeting of the Academy of Management, August 8–13, 2008. Reported by Christie Nicholson in a September 29, 2008, *Scientific American* podcast (scientificamerican.com/podcast/episode.cfm?id=business-lies-and-e-mail-08-09-29).
2. Sun was acquired in a fire sale by Oracle not long after this.
3. A *hash* is a computed ID based on the characteristics of an image or other file. Hashes are used to compare and sort large numbers of files quickly.

Chapter 3

1. Only the first message in a chain can be considered, since any Reply All command preserves the initial order.

2. While it is possible for companies to prevent this practice, they can do so only by essentially limiting all copying of data, which isn't practical. Instead, most companies content themselves with tracking only very sensitive types of data, such as trade secrets, strategic plans, and so on.

Chapter 5

1. Posted by "hauk" in June 2009 at reddit.com/r/program ming/comments/8u5dp/erik_naggum_19652009_rip.
2. Most Internet sources indicate that Erik had ulcerative colitis.

Chapter 6

1. As quoted in the online edition of *The State*, "Exclusive: Read e-mails between Sanford, woman: Sanford-Maria e-mails shed light on governor's affair," June 25, 2009 (thestate. com/2009/06/25/839350/exclusive-read-e-mails-between .html).
2. As quoted in *Grazia* magazine, Italy, October 2009. Research reported on page 6 of "2006 Workplace Romance Poll Findings, Society for Human Resource Management and CareerJournal.com," Society for Human Resource Management, January 2006 (shrm.org/Research/Survey Findings/Articles/Documents/06-WorkplaceRomance PollFindings%20(2).pdf).
3. This may be a cultural issue and therefore may not hold true outside the United States.
4. According to Wikipedia, *cyberstalking* is the use of the Internet or other electronic means to stalk, hurt, or embarrass someone. Wikipedia notes that typically the term *cyberstalking* applies to adults perpetrating the offending behavior on other adults, whereas *cyberbullying* requires

the target to be a minor (wikipedia.org/wiki/Cyber-bully ing#Cyber-bullying_vs._cyber-stalking).
5. "Man killed wife in Facebook row," October 17, 2008 (news.bbc.co.uk/2/hi/7676285.stm).
6. Katie Dawson, PA, "Life for wife-killer enraged by Facebook change," January 23, 2009 (http://www.independent.co.uk/news/uk/crime/life-for-wifekiller-enraged-by-facebook-change-1514002.html).
7. Liz Hull, "Husband 'murdered wife before killing himself' after she confessed on Facebook she was leaving him," May 4, 2008 (http://www.dailymail.co.uk/news/article-563 934/Husband-murdered-wife-killing-confessed-Face book-leaving-him.html).

Chapter 7

1. According to a study reported by the company Proofpoint on August 10, 2009 (http://www.proofpoint.com/news-and-events/press-releases/pressdetail.php?Press ReleaseID=245).
2. Serena Ng and Carrick Mollenkamp, "In UBS Case, Emails Show CDO Worries," September 11, 2009 (http://online.wsj.com/article/SB125262701573801493.html).
3. Note that laws differ greatly by country and even by state.

Chapter 8

1. See the excellent article by Max Chafkin, "How to Kill a Great Idea!" in *Inc.*, June 1, 2007 (http://www.inc.com/magazine/20070601/features-how-to-kill-a-great-idea.html.
2. One often-cited reason for this phenomenon is that the Philippines is a former American colony, so many peo-

ple speak English in addition to Tagalog, the indigenous language.

3. While the amount was undisclosed, the selling price was rumored to be as high as $100 million. In the years before the sale, the company had taken in almost $50 million in investment (http://en.wikipilipinas.org/index .php?title=Friendster.com).

4. Other social networking sites have also fallen prey to somewhat similar phenomena, such as the disproportionate popularity of Google's Orkut in Brazil, a country where people apparently change residences frequently and find the site especially handy for keeping in touch.

5. These sites generally ask whether you are a current or former employee of the company in question, but, of course, they have no way of verifying the information you provide.

6. Many tools can assess people's opinion of specific brands online, but these are mostly geared toward product lines rather than being about the character of the corporation itself.

7. Pete Carey, "Entangled: SEC Investigation of New York Hedge Fund Galleon also Ensnares the Fund Manager's Network of Silicon Valley Friends, Associates," December 6, 2009 (http://nl.newsbank.com/nl-search/we/Archives? p_action=doc&p_docid=12C77877011150C0&p_docnum =10&s_accountid=AC0110031023420528594&s _orderid=NB0110031023395628162&s_dlid=DL01100 31023421528632&s_ecproduct=DOC&s_ecprodtype= &s_trackval=&s_siteloc=&s_referrer=&s_username =cataphora&s_accountid=AC0110031023420528594&s _upgradeable=no).

Index

Abrams, Jonathan, 222
Amesefe, Eli, 5
Anger, 20
Animated images, 118
Anonymity, 2, 27, 187, 188, 234–35
ASCII copies, 120
Attachments (e-mail), 70
Attorney-client privilege, 12, 203–4
Automated employee performance analysis, 56–61

Bear Stearns, 46
Behavior, 19–20
Black English, 74
Blame, 103. *See also* Passing the buck
Blawie, John F., 200
Blogs, 98, 125–27
Blow Off Scoreboard report, 9, 10, 81, 82, 144
Body language, 142
Breakup blogs, 125–27
Buck passing, 100–109, 144, 215
Buck Passing view (Digital Mirror), 108, 109
Bull in the digital china shop (digital archetype), 131–37
Burton, Jim, 73

Capitalization (text), 50, 118
Cataphora
development of visual analytics by, 48
employees of, 73
and privacy issues, 191
role of, in D.C. Madam case, 185–86
and text, 57, 58
type of work done at, 5, 13

Cell phones, 74–76, 161
Centrality, 29–31
Chanel, Coco, 91
Character, 15–16, 100
Chat groups, 122
Cheating, 21
Cioffi, Ralph, 46–48
Class action lawsuits, 62, 193
Commentators, 28
Commenting, 86–87
Commingled communication, 132, 198
Communication. *See also specific forms of communication*
commingled, 132, 198
face-to-face, 115–17, 210
formality of, 76–77
intimacy of, 80
monitoring of, 205
multichannel, 173
software for monitoring, 240
styles of, 145–46
verbal, 166
Communication media, 79–80. *See also specific media*
Company policies, 205, 234–35
Compliance systems, 43–44
Computer programs, 56–61
Conference calls, 76
Confrontation, 54
Content creators, 59
Content curators, 59, 60
Context, 19–20
Contradiction, 46–48, 50, 199
"Corporate archaeology," 5, 15
Corporate culture, 54–55

Credit grabber (digital archetype), 102, 128–31
CyberAngels, 177
Cyberbullying, 177
Cyberstalking, 177, 180

Data, 231
Data records, 11, 61–63
Data visualization, 9, 202. *See also specific visualizations*
Data-entry order, 66
Dating services, 182–83, 185
D.C. Madam. *See* Palfrey, Deborah Jeane
Deadlines, 120
De-anonymization software, 188–89
Decision making, 103–5
Delaying language, 143–44
Delegation, 107
Dialect, 73–74
Digital archetypes, 97–139
bull in the digital china shop, 131–37
changes in, 143
created on social networking sites, 23
digital buck passer, 100–109
digital credit grabber, 128–31
digital exhibitionist, 122–28
last-word getter, 137–39
loud talker, 117–21
love bomber, 110–15
maintenance of, 91–92
and preference for face-to-face communication, 115–17
public shamer, 121–22
Digital buck passer (digital archetype), 100–109

Digital credit grabber
(digital archetype),
128–31
Digital exhibitionist
(digital archetype),
122–28
Digital identity integrity,
35
Digital Mirror software
Blow Off Scoreboard, 9,
10, 81, 82, 144
Buck passing view, 108,
109
delaying language
detection, 143–44
digital Pecking Order
assessment, 67, 68
downloading, 66
function of, 238–39, 242
graphs, 56
Loud Talking matrix,
89, 90
and loud-talker
archetype, 121
Quality Time chart,
93–94
Social You-niverse view,
220, 221
Stressful Topics matrix,
49, 83, 84
Temperature
Gauge meter
representation, 24,
158, 159
Who? What? When?
matrix, 85
Digital Pecking Order
assessment, 67, 68
Digital pillow talk, 171
Digital YOU
defined, 12
real you vs., 21, 22, 35, 51
Discrimination, 62, 193,
195, 205
Dissent, 54
Documents, 14, 18, 42
Draft e-mails, 70–71

E-mail(s)
attachments to, 70
and concealing personal
business, 23
credit grabbing in, 131

feedback on, 70–71
forwarding, 100–102
group, 122, 135–36
intimacy of, 79
length of, 81
loud talking in, 89,
118–19
and multichannel
communication, 113
order of responses to,
80–81
quoting in, 72
recipients of, 66–68
responding to meeting
requests with, 72–73
saving vs. deleting,
69–70
subpoenas for, 192
use of, while away from
work, 74–75
vernacular in, 73–74
volume of responses in,
93–95
E-mail threads, 114, 133
Emoticons, 49, 83, 84
Employers
communication
monitoring by,
205–7
and digital stalkers, 112
and public shamers, 122
and termination of
employees, 17
Emurse, 7–8
Exhibitionists, 122–28
Expertise, 31, 32, 216

Facebook
and digital
exhibitionists, 127
job applicant
information from, 2
levels of interaction
on, 87
personal content on, 173
relationship status on,
169, 177–79
Face-to-face
communication,
115–17, 210
Fantasy sites, 227
Feedback, 98, 215
50 Cent, 39–40

Financial sector, 234
Flickr, 2
Flirtation, 166, 168. See
also Romantic
relationships
Flow of Information chart,
56
Fonts, 118
Foreign talking, 87–88
Formality, 76–77
Forrester, Wayne, 178
Forums, 122, 131, 137,
230–31
Forwarding e-mail, 100–102
Fraud investigations, 16–17,
152, 154
Freeze letters, 202–3
Friendship, 222–24
Friendster, 217–19,
222–23

Gaming sites, 227
Generation gaps, 115
Glassdoor.com, 27, 228
Gmail, 197, 207
Google
and digital footprint, 3
function of, 32–33
Googling yourself, 39
job applicant
information from, 2
ubiquity of, 34
Grammar, 78–79, 87, 99
Gran Torino (film), 167
Graphs, 29–31, 53–54, 56
Grazia (magazine), 168
Grinhaff, Tracey, 179
Group e-mail, 122, 135–36

Hashes, 42
Heraclitus, 15
Hidden organization effect,
106
Hobbyists, 83
Holidays, 75
Home pages, 119
Human resources
departments, 18,
213–14

Illegal activity, 205, 211
IM. See Instant messaging
Insider trading, 234

Instant messaging (IM)
and concealing personal
business, 23
content discussed in, 93
formality in, 79
inappropriate, 208–9
loud talking with, 119
during meetings, 75–76
Insurance, 204
Internet, 28, 128–30
Intimacy, 79–80
Invitations, 72–73, 86
IP addresses, 137, 218, 219
IT departments, 69, 201

Jackson, Curtis, 39–40
Job seekers, 105–6, 238

Klingon, 41
Knowledge, 231

Language, 41–42, 143–44,
174
Last-word getter (digital
archetype), 137–39
Lawsuits, 7
class action, 62, 193
collection of personal
data for, 61–63,
192–95
from contradictions,
46–48
and industry sectors,
212–13
process of, 201–2
steps to take when
anticipating,
202–5
UBS lawsuit, 200
Leadership, 215
Legal departments, 18
Leilani de Jesus, Carmen,
218
Linguistic fingerprint
analysis, 138
LinkedIn, 2, 34, 39
Loud talker (digital
archetype), 89, 117–21
Loud Talking matrix,
89, 90
Love bomber (digital
archetype), 110–15
Lunt, Chris, 218

Lying, 21
Lyndon, Roger, 168

Managers, 103, 155, 211,
213–15, 240
Massively multiplayer
online role-playing
games (MMORPGs),
227
Matt-welsh.com, 35–36
Mattwelsh.com, 35
Meetings, 72–73, 75–76
Meter representation,
24, 25
Microsoft Office
documents, 42
MMORPGs (massively
multiplayer online
role-playing games),
227
MOL, 219
Monitoring
of employee
communication,
205–7
risk management, 41, 212
software for, 240
Moods, 50
Multichannel
communication, 173
Multimedia presentations,
14
Multitasking, 115, 117
MySpace, 127

Naggum, Erik, 147–52
Newscorp, 217
N-grams, 57, 58

"Offliners," 115–17
Online dating services,
182–83, 185
Online receipts, 197
Opinion, 24, 26, 199–200
Organizational change, 53
Outlook data, 95

Palfrey, Deborah Jeane, 4,
185–87
Passing the buck, 100–109,
144, 215
Performance analysis,
56–61

Personal digital brand
management, 34–35
Personal information
and breakup blogs,
125–27
concealing, 23
of other people, 88, 134
Personality traits, 6. See
also Digital archetypes
Philippines, 218–19
Phone calls, 74–76, 113, 115
Phone records, 93
Picture of Dorian Gray, The
(Oscar Wilde), 23
Pictures, 92
Pornography, 42–43, 63,
136, 205
Power accumulators, 103,
106, 108
Privacy, 110, 191, 197, 209–10
Privilege log, 204
Productivity, 56, 59–61, 155
Prosody, 167
Public relations, 231
Public shamer (digital
archetype), 121–22
Punctuation, 50

Quality Time chart, 93–94
Quoting (e-mail), 72

Racial epithets, 205
Rajaratnam, Raj, 234
Rejection, 225–26
Responsibility, 102
Résumés, 2, 7–8, 105–6, 196
Richardson, Edward, 178
Risk management
monitoring, 41, 212
Romantic relationships,
161–89, 239
appropriate workplace,
170–72
and breakup blogs,
125–27
communication cues in,
167–68
ending, 174–80
identifying existence of,
164–66
and online dating
services, 182–83, 185
and sexting, 181–82

and sexual harassment
    cases, 206
and social media, 169–70
sustaining contact in,
    162–63

Sanford, Mark, 162–63
Schon, Keith, 5
Schwartz, Jonathan,
    27–28
Sensitivity, 99–100
Sentence structure
    analysis, 50
Sexism, 206
Sexting, 181–82
Sexual harassment, 206,
    208
Sexual relationships, 165
Shadow networks, 233
Skype, 21, 110, 140, 175, 186
Slander, 180–81
Social invitations, 72–73,
    86
Social networking sites.
    See also specific sites
    access to content on, 127
    friendship on, 222–24
    interconnectivity with,
        226–27
    personas created on, 23
    posting as somebody else
        on, 34
    reality vs., 219–20
    and rejection, 225–26
    and romantic
        relationships,
        169–70, 181
    uploading pictures and
        videos to, 92
Social networks, 29–31,
    53–54, 86
Social proximity, 42, 93

Social You-niverse view
    (Digital Mirror),
    220, 221
Software
Digital Mirror. See
    Digital Mirror
    software
    and knowledge, 231
    for monitoring
        communication,
        240
    for risk management,
        41, 212
    for scanning résumés,
        196
    for spelling correction,
        78
    for tracking intimacy of
        communication, 80
Something's Gotta Give
    (film), 169
Speed-dial list, 67
Spelling, 78–79
Stalking, 110–15, 177, 180
Stealing, 20, 21
Stressful Topics matrix, 49,
    83, 84
Subject headers, 119
Subpoenas, 186–87, 192,
    194, 195, 201
Sun Microsystems, 27–28

Tannin, Matthew, 46–48
Temperature Gauge view
    (visualization), 25,
    158, 159
Termination, employee,
    17, 42
Text, 57, 58, 118, 120
Text messaging, 181–82
Theft, 20, 21
Time, 93–95

To Have and Have Not
    (film), 167–68
Tobias, Randall, 186
Tone, 146
Topic drift, 133
Truth, 62
Turnaround times, 145
Twitter, 86, 173

UBS, 200
Up in the Air (film), 116

Verbal communication,
    166
Vernacular, 73–74
Video clips, 92
Visual analytics, 48, 49
Vitter, David, 186
Vocabulary analysis, 50
Voice mails, 80, 113

Weekends, 75
Welsh, Matt (Harvard
    professor and software
    engineer), 35–38
Welsh, Matt (male
    model and software
    engineer), 35–38
Whistle-blower cases,
    153–54
Who? What? When?
    matrix, 85
Wikis, 122, 131, 135
Wilde, Oscar, 23
"Wisdom of the collective"
    notion, 26
Woods, Tiger, 162
Written communication,
    166

Yahoo! message boards,
    137, 230

DATE DUE